TREATISE O

TECONOMICS

OF

DYNAMIC RISKS

ALL NATURAL DISASTERS,

&

ENERGY RESOURCES PRODCUTION DISASTERS

Bahman Fakhraie, Ph.D.

TECONOMICS OF DYNAMIC RISKS

All natural Disasters, And Energy Resource Production Disasters and Some Financial Risks

By

Bahman Fakhraie, PhD

4/2/2013

$28.95

ISBN 978-0-9852958-5-1

52895>

9 780985 295851

Acknowledgments

Many thanks to my family for they have been the many joys in my life. Mrs. Kay Davis Fakhraie received her master in Speech Pathology from Utah State University. Daughter, Lara Fatemeh Fakhraie, earned her Master's degree from Oregon State University. She is a published author. She resides in Oregon. Son, Anayat Fakhraie, a screenwriter earned his Master from the American Film Institute. They did their undergraduate work at University of Utah. Fakhraie families have endured revolutions, death of family members, and hardships. The sui generis caravan of joy de vive continues. I wish for all of the family members, professors, teachers, students, helpers, and friends spread all over the globe the best, warm thoughts, and many thanks. Life events are neither all fair nor happy. I wish they would always leave you better off than before.

Abstract

All production processes have some inherent risk.
An important part of innovation development of new
process of technological innovations was learning to master
the exceptions to the rules, which reduces risks. Therefore
and important part of gestalt of new technologies
injections, and new variables of modular production
processes proceeding almost all thought-action processes
introduce by Dr. Bahman Fakhraie in his studies were the
endogenous elements of production process and the new
variables, including the educational aspect of skills among
others, which would reduce some risks. Otherwise
humanity, would have not stepped out of the caves because
of risks in becoming fast food for dinosaurs, large cats, and
predators. The magic difference among the mammals and
other species are the capabilities to learn in group defense,
or individual survival tricks and skills and retain that

knowledge, and pass on those skills to improve the chance for their offspring survivals, for the genome continuations.

The occurrences of natural disasters (floods, earthquakes, hurricanes, cyclones, etc.) are rarer events; however, they do reappear on longer-term cyclical patterns, sometime in the same regions.

Contents

Table of Figures

Preface

Dr. Bahman Fakhraie, PhD in Economics, University of Utah, his dissertation added to the influences of the Unorthodox Holistic Economic doctrine and complemented the modern Orthodox economic theories, in the millennial age of technological paradigm shifts. He applies analytical skills with gestalt study of history, mathematics, and econometrics to economic analysis, with scientific background. He is a Published Economist, Author, Researcher, Investor, and Private Contractor. His skills are in international trade and finance, economic production (theory and application), growth and development theory, econometrics, agriculture economics, and agronomy. These are greatly valued skills combinations to employ. Dr. Bahman Fakhraie was invited by Senator Frank M. Browning to attend Utah State

University, when Senator visited his family. He has met most of modern day science and economic opinion leaders. Dr. Bahman Fakhraie's book web page is at following link, http://bahfecon.wix.com/bahfecon.

This book is the result of many years of notes and journals, which Dr. Bahman Fakhraie wrote after different disasters, since Tehran, Iran earthquake. San Francisco Earthquake of 1980s, and hurricane Kathrin, South East Asian tsunami, and Japanese, earthquake, tsunami, and nuclear disasters all combined.

CHPTER 1

Treatise on Teconmics of Dynamic Risks

All Disasters, And Energy Resource
Production Disasters

The process of living is a continuum. Life on terra

firma is also dynamic continual process, a natural dynamic

production process, which entail probable riskiness on its

path of dynamic but continuous alterations, experienced

and or caused at the dynamic modular production levels.

Therefore, there are uncertainties associated with

the atomized modular production processes, which

accumulate to lifelong experiences. Hence, these studies of

risk associated with modular production processes are not

hostile or attacks on body or mind of human knowledge,

but essential components of advancement of human capital

skills. In the same category of important technological

injections of human capitals, as education and research of

financial risks, which enhance returns, profits, and rewards of modular productive investments, and reduce chances and occurrences of risk associated with such endeavors.

The occurrences of natural disasters (earthquakes, hurricanes, cyclones, floods, etc.) are rarer events; however, they do reappear on longer-term cyclical patterns, sometime in the same regions. From the tectonic plates movements to costal earthquakes, tsunamis, as human knowledge of observations have evolved in details, the metrologies of survival have also advanced gradually.

Introduction

The gestalt of new technologies injections and new variables of all modular production processes can continue, after almost all of the essential initial thought to action sequences of productions.[1] Dr. Bahman Fakhraie introduced in his studies the endogenous elements of production process and the new variables that were the educational aspect of skills, which among other things they would reduce some risks.[2] Otherwise, humanity first thought-action modular production process would have

halted. That is, the first hominids would have not stepped

out of the caves because of risks in becoming fast food for

dinosaurs, large cats, and predators.

It is a truism that all disasters, and catastrophes, like

politics are local.[3] However, the frequency of occurrences,

their global scopes, the size of damages to human and

nonhuman capital, and duration of recovery defines them.[4]

However, their magnitudes combined with hyperactive

(uber) multimedia communicative connectivity make them

a study subject of essential science phenomenon.[5]

Therefore, this study focuses on major disasters and event

that are damaging economically at macroeconomic and

global levels.

President John F. Kennedy in one of his memorable

speech firmly stated, "All these disasters are people made,

therefore people can fix them." There have been some

publications, and organizational time and efforts spent on

this task. FEMA[6], USDG, department of homeland

security in United States, Red Cross, and Coast Guards and naval task forces combined with regional and local military and civilian government, and local NGOs (none governmental organizations) assistance and cooperation have produced more specific and general papers, notes, suggestions, booklets, and books about different aspects of this subject. More importantly, they have saved lives and helped people in despair.

This area of knowledge has become a much more dynamic field of study after the increases in natural phenomenon of extreme climatic changes. This study will only show reasons that such informational and governmental resources and efforts are required dynamically and in the longer run.

The following study reviews some scientific studies, with added personal notes and experiences during or after each disaster focuses on the essentials tasks reminders, which we forget between the disasters.

Since, these fields is fresh and dynamically changing, and unpredictable, this study is by no mean exhaustive, nor any study can be at this stage of knowledge accumulations. Therefore, this is a testimonial to the needs for the study, and that fact that no guarantees are given legal, academically, or otherwise.

Natural History

The scientific origin of tectonics plates movements are not difficult to comprehend, now that visual evidences exist, however, the scientific study of issue under dynamic tectonic plate studies is very new science. Figure1shows a selected presentation of tectonic plate movements, from 250 million years ago to present, which also represents the continental drifts graphically. This field of study was a well-fought battle of dimwittedness for some time, even the scientific venture into the fields 1950s-1980s were not sure footed.[7] Figure 1 shows the dynamic nature of earth contents' surfaces, which come in contacts and remain in motion constantly.

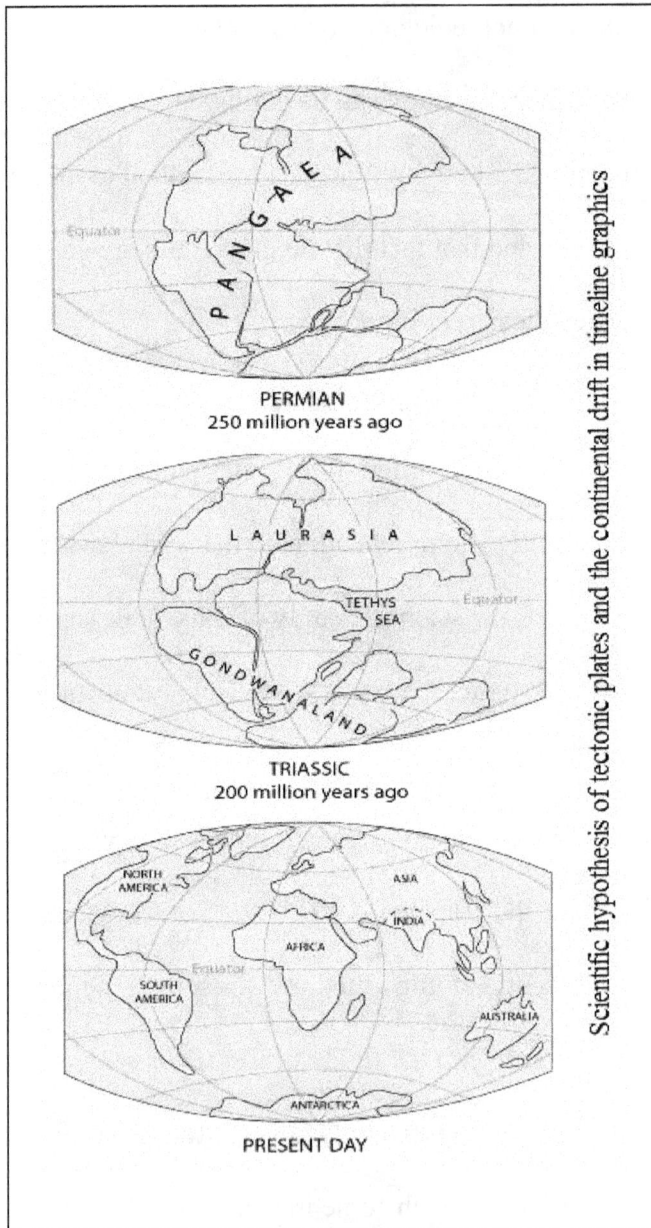

Figure 1, The dynamic nature of earth tectonic plates.

That constant motion will create energy and forces over time that causes plate movement, earthquakes, and volcanic releases activities.[8]

Nevertheless, the dynamic tectonic plate studies are evolving quickly under scientific methodology and contributions of many specialized fields of studies, like this study. Planetology, seismology, geography, anthropology, botany, oceanography, microbiology, and biology, climatology, hazard science, data analysis and econometrics are just a few contributory fields.

Since, the hard work, observations measurements and hypothesizing, testing, and theorizing have been accomplished. There are fantastic underwater videos, multimedia, films, about separations and dynamic movements of different parcels of earth, thus the definitional titles of the tectonics plate's movements.

For millenniums, humanoids were intimidated from elements, lightening, thunder, fire, earthquakes, and cave-ins. It is only recently in less than two millenniums, a few hundred years that fear gave to curiosity, and searching

curiosity to knowledge seeking about the same elements.

That is extensive studies of natural phenomenon became

internalized to the process of discovery and accumulation

of human knowledge.

Natural Disasters

It helps to know what risks are associated with

warnings, alerts, and government statements, over time.

The next graph, figure 2 shows that the earth moves

apart from the middle of the ocean and is pushed under the

continental plates at the west coasts of USA. American

continents' hotspots are designated from west coasts of

Alaska, USA, and South America. Eurasians' continents

western fronts stretch from Italy, to Iran, Afghan and

Pakistan, and India.

Figure 2 shows global seismic hazards map.

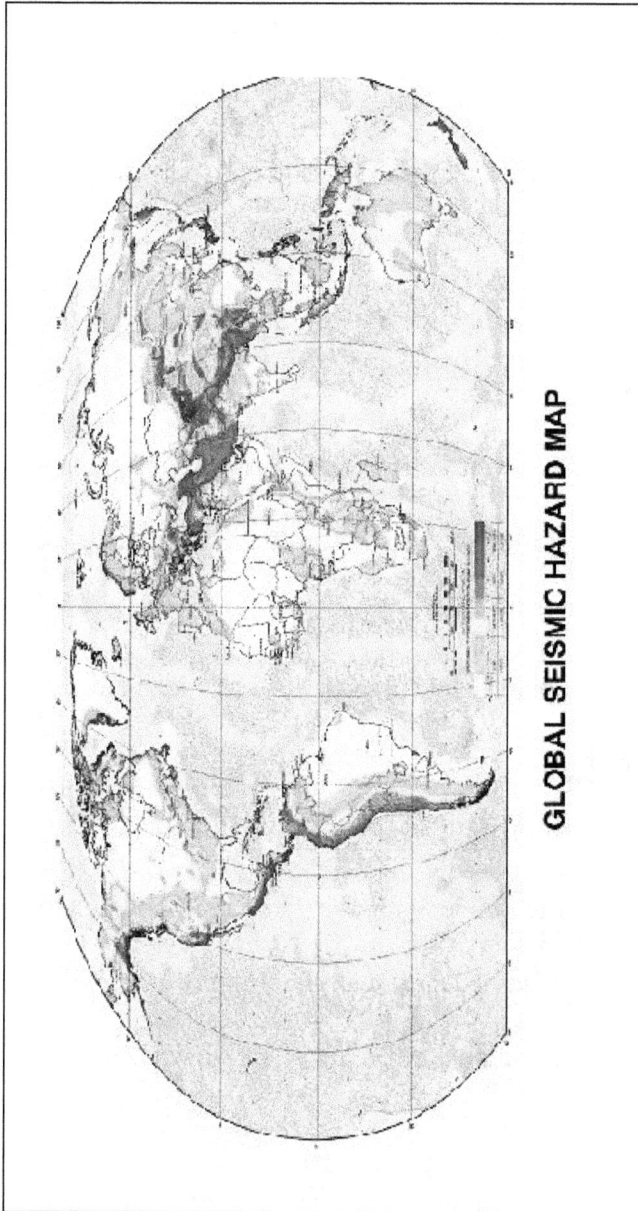

Figure 2, Global seismic hazards map.

Those countries are all witnesses to horrendous events, with mass collaterals and damages. Even more recent, Japan, earthquake, Tsunami, and industrial disasters had natural bases. Nevertheless, China, Korea(s), Southeast Asiatic nations have all been victims of such occasional disasters.

In the east, Iran, to India experience the same level seismic risks of earthquakes, from such movements as well. The higher the risk the warmer the colors, including color red indicate increased probability of seismic events, associated with continental drifts.

Probabilities and Costs Estimations

The public often under estimate them until too late. For example, in cases of earthquakes, falling debris associated with cave-ins can be harmful and fatal. In case of a tsunami, the wall of water with debris can over run buildings, cars, or individuals. In cases of nuclear and petrochemicals disasters, toxins, toxic fumes, radioactive particles, visible or not can be fatal, and remain in the immediate living environments for some times. Therefore,

duration, proximity, location, power, or weakness indices

of the disasters can help,

- It will rationalize a plan of action,

- It will qualify and to quantify the risk magnitude,

- It will help reduce fear in some cases,

- It will help galvanize a plan of action,

- If possible; therefore, those plans of actions, for

  escape or survival can be organized, or formulated.

- In this case A, there is positive chance for dynamic

  risk, $\frac{dr}{dt} > 0$

  1-In cases of service, good, or investment qi, there

  is a risk-chance Pqi can increase or decrease,

  subject to probabilities (sd = δ, $\pm$).

  2-π = f (p, q, cq, rq) = f (+bq, - cq, - rq) = (p. q - c)

- In this case B, there is zero or negative chance of

  dynamic risk. $\frac{dr}{dt} \leq 0$

With a concern eye on landing of hurricane Sandy, in North Easter of Unite States, from Atlanta to New Jersey, New York to Canada. East coast had hurricane Irene recently, and Canada had bicoastal twins of an earthquake on the west and the looming hurricane Sandy on its east in October 2012. It is important to dissect the dynamic teconomic risks of these disasters.

The magnitudes of Japanese multiple calamity defies verbal mastery of mere mortals beyond condolences for the toll of it like other disasters.[9] The economic and political debris and damages of hurricane Katarina for individuals, and some states are still there. Despite the glory advertisement spins associated with certain oil productions, the full toll and implications of deep-water oil well disaster in food industries, fishing fleet disasters, and environmental long-term implication will be with us for some time some invisible. Some of it only washes ashore like wades of gummy oil patches visually, or with bodies of diseased fish with unknown illnesses, yet undiscovered.

While the tourists may be lured back early, some small businesses and fishing operations, shrimping outfits, clam and crawfish shacks have closed and are gone for good.

Figure 3 shows estimates shows level of funding to study and monitor such event, in recognition of their importance and the human lives they affect.

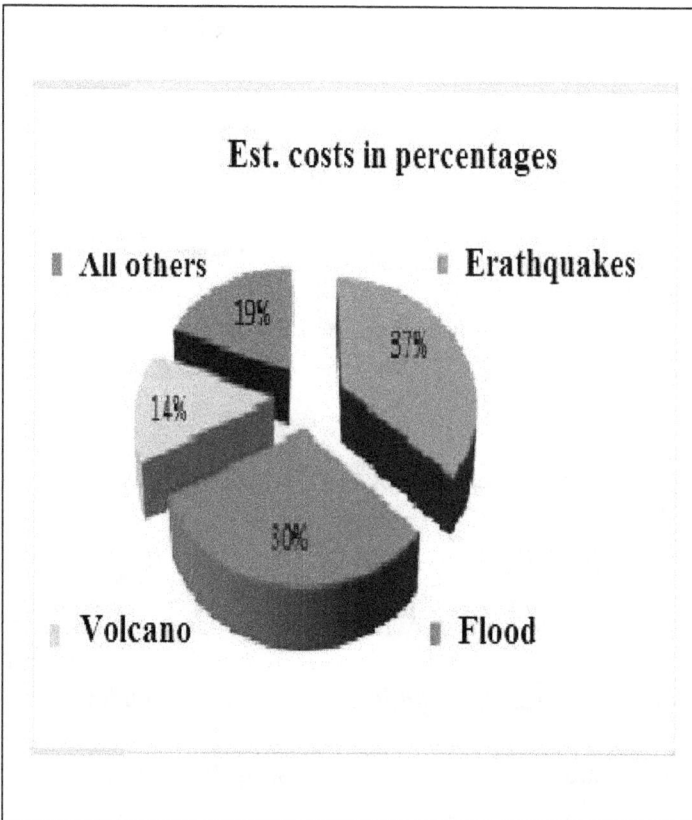

Figure 3, Estimated hazard funding reimbursements

This is an approximate U.S. Geological Survey appropriated and reimbursable hazard funding levels.[10] Author has recalculated the costs for this this graph in est. percentages pie chart.

There are a few good reasons industrial natural disaster combination will remain a concerns,

- Population numbers and density is increasing globally.

- The demands for industrial energy and technological injection have increased their numbers and their approximation to population centers.

- The job- creation aspect of supply-side, without increase in government-induced investment in renewable energy will exasperate these situations.

Therefore, that should not confuse the reader, the Japanese natural industrials disasters, or the American Mexican golf deep-oil well disaster or future nuclear disasters can dwarf the data, the negative externality potentials and probabilities do not serve humanity well, if events are dismissed them lightly.

Figure 4 shows as the percentage of disaster data collected. It serves as a reality check, when we compare some of the more recent major industrial disasters with the nature disasters.

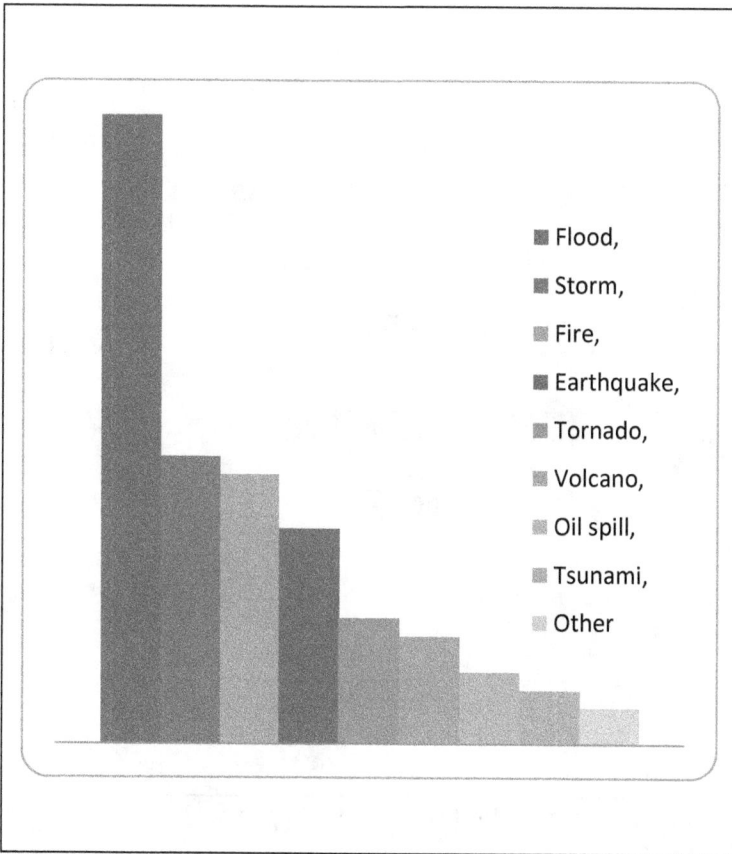

Figure 4, Emergency event data in percentages.

Figure 5 shows number of people killed by disaster types.[11] These are 2009 disaster numbers as presented by the sources.

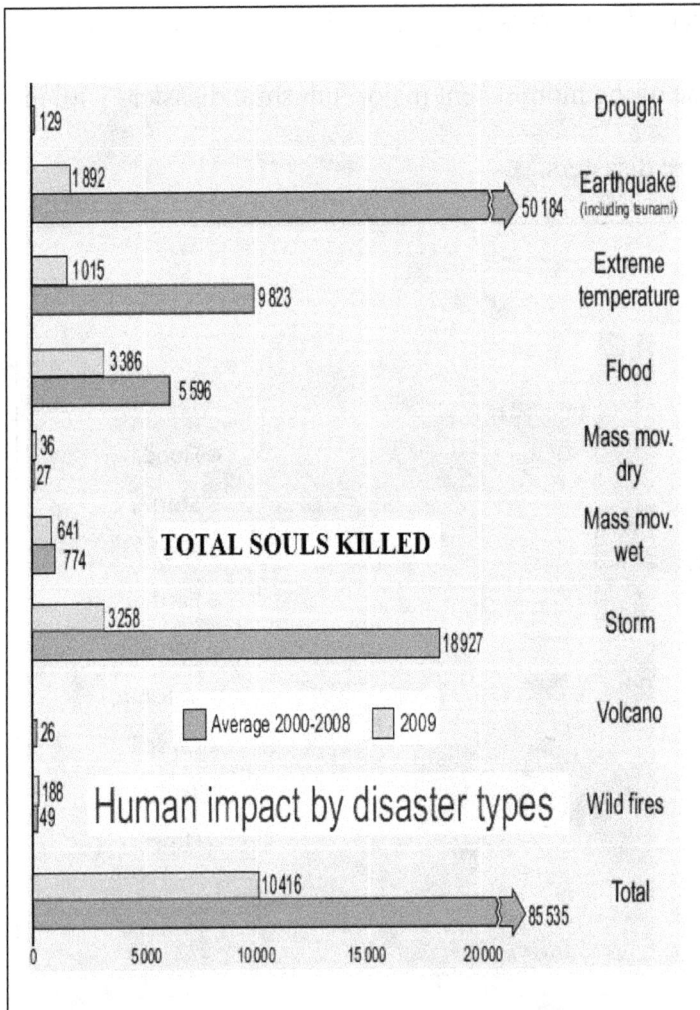

Figure 5, Human killed by disaster types.

The following two graphic representations demonstrate amply why the concerns have heightened in these areas of research and studies. Moreover, resources should go to study focused research like this study.

Figure 6 shows annual reported for economic damages, 1980-2011 over time, in this graph trend lines estimation are diverging to the upside.

Figure 7 shows the trend line for reported numbers of disasters are increasing. Therefore, it also reconfirms that (figure 6) disaster's economic-damages trends are increasing over time.

In case, this is all cyclical, or data collections have improved substantially, the data still covers 30 years, way beyond any economic long cycles, of 3,5,11 years. It will serve logic well to be prudent about alternative technological injection solutions.

Figure 6, Economic Trends of damages from disasters, (1980-2011).

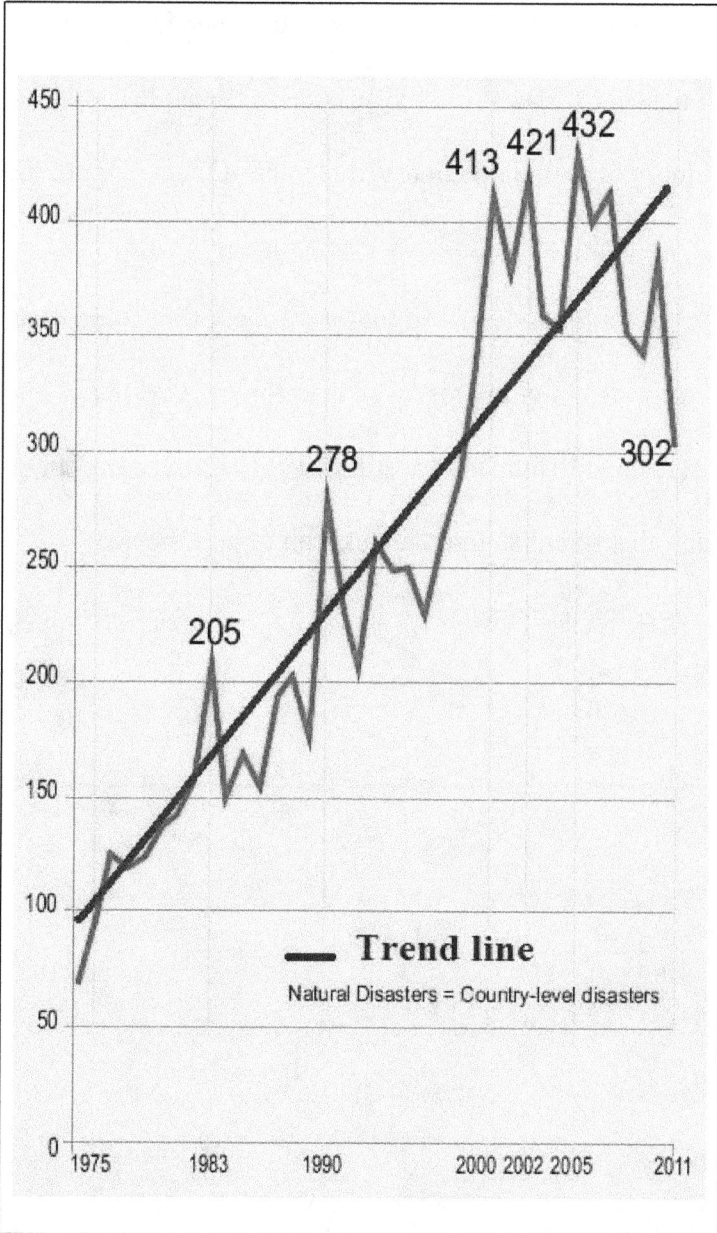

Figure 7, Reported natural disasters trends (1975-2011).

However, a large number of commercial experts, political show maestros, and science salespeople are all out spinning for some imaginary outcomes. Often, it is not too early to blame or sell, rather than assist!

Yet, to assist in the future parallel events, one needs to study such nightmares, as they unfold pictures by pictures, while humans are still in morning and pain. In order, that scientist and planners can reduce the heavy losses of such disasters.

Notes

[1] *Teconomics: the microeconomic analysis*, Utah, FERDAT
publishing 2012, and,
https://www.createspace.com/4196760, Books web page at,
http://bahfecon.wix.com/bahfecon

[2] *Technological injection, dynamic new capital
measurements, and Production Theory in Economics*,
(Michigan: ProQuest LLC, 2010) and,
https://order.proquest.com/OA_HTML/pqdtibeCCtpItmDsp
Rte.jsp
Books web page at, http://bahfecon.wix.com/bahfecon

[3] Dr. Bahman Fakhraie, *"Treatise on Dynamic Risks, All
Disasters, And Energy Resource Production Disasters,"*
scientific paper, 10,28,2012,
https://plus.google.com/u/0/?tab=mX#stream/circles/p57ebfa130
e9cc418

[4] Holmes, R.R., Jr., Jones, L.M., Eidenshink, J.C., Godt,
J.W., Kirby, S.H., Love, J.J., Neal, C.A., Plant, N.G.,
Plunkett, M.L., Weaver, C.S., Wein, Anne, and Perry, S.C.,
2012, Natural hazards science strategy: U.S. Geological
Survey Open-File Report 2012–1088, 53

[5] For more definitions and some classification of disasters,
"Guha-Sapir D, Vos F, Below R, with Ponserre S. *Annual
Disaster Statistical Review 2010: The
Numbers and Trends*. Brussels: CRED; 2011." p.9
"Encourage the free use of the contents of this report
with" appropriate and full citation:
The data upon which this report is based are maintained
through support of the US Agency for International

Development's Office of Foreign Disaster Assistance (USAID/OFDA),
http://www.emdat.be/classification,
Centre for Research on the Epidemiology of Disasters (CRED)
Université catholique de Louvain – Brussels, Belgium,
This document is available on
http://www.cred.be/sites/default/files/ADSR_2010.pdf.

[6] FEMA: Federal Emergency Management Agency,
USGS: United States Geological Survey,
NEIC National Earthquake Information Center
NHMA Natural Hazards Mission Area
NISAC National Infrastructure Simulation and Analysis Center, NOAA National Oceanic and Aeronautics Administration,
NSF National Science Foundation
NSIP National Streamflow Information System
NVEWS National Volcano Early Warning System
SAFRR Science Application for Risk Reduction Project
GSN Global Seismic Network, H-SSPT Natural Hazards Science Strategy Planning Team, NASA National Aeronautics and Space Administration, Ibid

[7] http://pubs.usgs.gov/gip/dynamic/historical.html

[8] Ibid

[9] RT,
Strong_8.9_earthquake_rocks_Japan_tsunami_hits_North_East

[10] Holmes, R.R., Jr., Jones, L.M., Eidenshink, J.C., Godt, J.W., Kirby, S.H., Love, J.J., Neal, C.A., Plant, N.G., Plunkett, M.L., Weaver, C.S., Wein, Anne, and Perry, S.C.,

2012, Natural hazards science strategy: U.S. Geological
Survey Open-File Report 2012–1088, 75 p.

[11] Data sources are: Center for Research on the
Epidemiology of disasters (CRED), Dept. of Public Health,
www.emdat.be, www.cred.be,
International Strategy for Disaster Reduction (UNISDR),
www.unisdr.org

CHPTER 2

Energy Resource Production Disasters

We have had chains of energy resource disasters with full emotional and economic impacts everywhere. Because, the demands of modern human lives and commerce operates with produced energy. These have been manufactured or people-made disasters. Therefore, some degree of human technological injections can control damages. It should help to have a cheat sheet ahead of these disasters to guide most human mortals in those choppy after effects storms.

Figure 8 shows actual and projected estimations, and a worst-case option for total energy consumption and production.

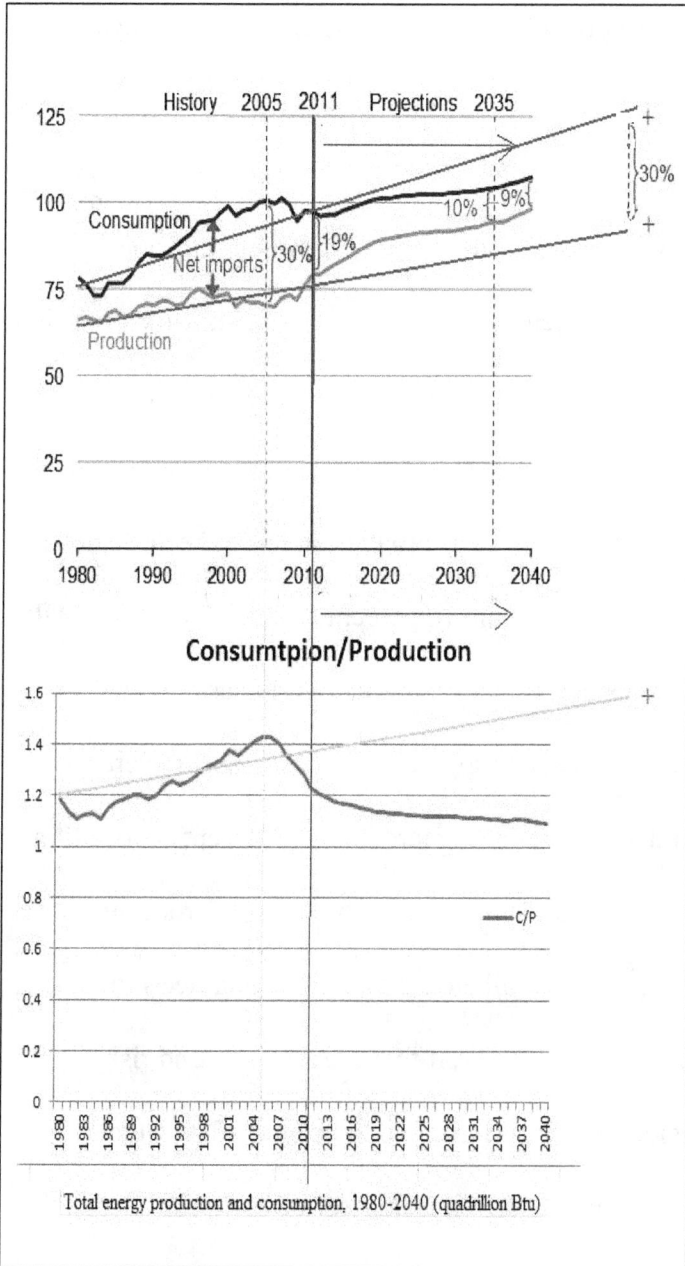

Figure 8, Total energy consumption and production estimates.

In projection-estimations, there are also always the probabilities that things are not going as planned.[1] The worst-case scenarios, where growth and consumptions of energy increases, and no strategic policy implementation are not presented or followed, those are shown by (+) signs. Imports will rise again exceeding the 30 percent.

Oil

The total energy production has to be sustained and increased to cover the 10 percent that consummation will exceed domestic energy production, in case of positive scenario. Total energy production also has to increase sustainably to cover to more than 30 percent shortfall, in the worst-case scenarios. The production processes in these cases, have all elements of risks and decisions making involved with them.[2] At those level risks, and risk eversions elements have had to be studied, weather they were included or not, it is another issue.

The BP-USA-Mexican-Golf deep-oil well spill, the Alaskan oil-tanker spill, and the overall damage to seafood industries and complement industries are significant. The higher risk associated to none sleeved multi-directional deep sea drillings has been barely noted, after the half multiple years of polluted seabed and of the seafood baskets of the West, in the Americas. The 20 billion dollars set aside legal fund, required for the first time, by President Barak H. Obama may be a giant leap from the past accommodating administrations and political operatives. Yet, the financial estimates could fall short of the actual accumulative long-term damages, hopefully not by too much.

Even a quick visual study of history of oil production around Texas, Caspian see, and Southern Russia, South America, and Africa, and globally, reveals volumes about pollution and toxicities, negative economic externalities about acreages of toxins associated with old and new production. The political aftereffect on localities, and their economic conditions are quiet another chapter of multiple disasters and political credulity.

The Alaskan oil tanker disaster left prolonged effects on the village lives of nearby population. Fish schools, and animals' food and wildlife numbers were affected in huge numbers. Major population shifts in schools of food fish influenced harvest, other dependent fish, and animals in the area. Some smaller villages were relocated to areas that were more hospitable, with further human and accumulated negative externalities and costs.

<u>Supply or Production, and Disasters</u>

There are a lot of losses, many death, grief, and depression after a major disaster, the emotions are raw, everyone needs to take time out for that grief, hunting for supply afterward will only add to it. Therefore, we all need to be patient with each other. Second day after the San Francisco earthquake in late 1980s, I was in a small polite queue, in front of an upscale Palo Alto grocery store, south west hills of Stanford University. The store had opened temporarily to take care of the real need and supply shortages of their consumers. They also needed to clear the shelves, some already cleared by the over 8-points

earthquake. They were nice, they stated take what you

need. They also said they would reopen later that week if

not the next few days, when the electricity was back up.

The electricity went back on in a few days. Water, bread,

frozen meat, can food for a week for three or four people

were available, which we shared with guests that night.

The stores were opened in a few days. Exceptions were a

few areas around the downed bridges. There were long

queues for gas, for coffee shops, to get in the grocery

stores, emergency rooms, and for medical assists. Besides,

the usual traffic around rush hours, which is normal in most

metropolitans. The main problem was not the production

of these items, goods, and services. The production had not

changed. It was the supply of what public needed, and

where they need it at that time.

Sciences will tell you. Almost all outer islands will

be under the one or two feet of oceanic water level rises,

because of climatic warming, and the ice caps that are

melting. You ask the scientist to show you. They show you DVD of American Alaskan tribal villages, which had to relocate. They show you the pacific islanders losing grounds. You say you do not believe it. So never mind, we are talking religions now, not scientific proof.

Nevertheless, here came Katrina, Irene, Sandy, pushing water on top of raised water level in New Jersey, and outer islands, and then lovely Manhattan, subway systems, Hoboken, and Stanton island, they were under 4 to 5 feet of salt water. Now, do you want to put a wall up from Louisiana to Canadian border? Do we want to study it? Maybe it is too much money to study it; Hurricane Sandy alone may cost an estimated $75 to $110 billion of dollars. Do you want to experience the next storm blind without sciences?

Prices, Queues, Rationing and Quantity Controls

High free market prices or price control, one causes class distinctions (wars) and biased-benefits to top few percent the very rich, the other causes shortages and inflationary pressures both, in the short run without

production changes. Disasters are another example, of free

market are unable to be perfect, or free. Even in normal

time, gas shortages have caused enough problems. There

are a few ideas developed to deal with them. We had

President Nixon price control days. Under President

Jimmy Carter, the economy faced gas shortages too.

Slower speed 55 mph versus the highway speed of 75-80

mph, and odd-even days driving and visiting gas stations,

requiring more than one person per vehicle, these all

worked, but not without usual complains. Since, a more

holistic unorthodox economic vision has shaped the smarter

economic mindsets. It has injected smarter technologies in

the mix economies, globally.

The economy is using multiple energy source

vehicles, and higher miles per gallon cars now. We are up

from 7-8 miles per gallons cars (1960 to 1970s, and 33

cents a gallon of gas), to 30-55 mpg cars, or electrical cars

(2012, and $ 3.60 in UT to $5.20 in CA, per gallon of gas).

The industries and manufacturing will also become educated in that way.

Those holistic unorthodox economic and technological injections have complemented the orthodox economic advancement well. Those technological injections will stay in modern economies, or there will not be a modern economy. Because the one supply energy source of production stratagem has changed, even in the energy producing and supplying countries multiple-technologies energy sources and uses are studied attentively. Some of these multiple-technologies energy sources other than petroleum bases have been adopted fastidiously, without declaration of wars on other cultures or countries, or screaming at the darkness or deities and others faith.

Even with preplanned strategies, great logistic relocation of supplies, there will always be glitches, mistakes, problems, after all disasters. Therefore, reasonable rationing by governmental entities closer to the problem with consultation with chain of command should be able to enact rationing and control of quantity, and quantity used for a temporary periods, until adjustment in delivery, relocation of supply can take place, 3-days to one week often is enough in advanced economies. However, these can change or need modifications in different circumstances, nations, localities, and economies.

Casting Votes, Queues, and Imperfect Free Markets

Two new examples follow about why perfect free markets cannot be perfect or free, after major disasters. Even though voting and disasters are uncomfortable as part of the same prose, here are the examples. One example is the presidential voting, in North East after hurricane Sandy. There will be elderly in high-rises without electricity, which cannot move out without assist to vote. Some voters relocated to different Red Cross centers or locations, they need to vote. There are voting-centers, which have no electricity or may be under water, or even washed out. There are four days to arrange for the presidential election voters. Let us see how many votes are casts in this region.

These supplies and demands adjustments count in a

democracy.

The second examples can be more problematic if

not insidious. That is, when long-queues form for long

hours, and voters cannot cast their votes in early votes, or

in Election Day, for whatever reasons. When that occurs in

the same general states with the same political

organizations repeatedly, there is more behind it than

ineptitudes of usual political cronies. After presidential

election of 2000 in United States, and the Florida and

national extended counts, there were real solutions to

remedy the issues. Early votes, absentee ballots, for up to

two weeks worked fine. The uses of computers with paper

proof of votes were to reduce the problems of long hours

and queues. However, late date experimentations close to

elections with both concepts have distracted from the

mission, besides chiming the legal alarm bells. That is

adding to the imperfections of elections and its free market place.

Queuing proves shortages of supply in a free market place, often a production process issue. However, in national presidential elections, that should not be a problem, ballets, or papers are plentiful. It is organizational, or for other institutional motivations, costs and volunteers, complexities of certain ballets and computers cause production process issues. Never the less, federalizations of such votes handled in certain states inefficiently, and repeatedly would remediate costs and inefficiencies, and relieve the long hours of queues very quickly. In time of disaster, national guards, or federal entities (volunteers) can relieve some of these duties, and issues, as they do now. The right to vote is still a right and not a privilege for all parties. Even though the constitution may have assigned it to states to decentralized it, if it is being mishandled and abused repeatedly, there should be remedies to improve its functionality.

In this age of computerization, the mailed or downloaded-ballots can be filled, and the finished product can be forwarded, mailed, taken in person, and be presented with all of those ID-requirements, which should not add to the costs of voting as an illegal toll-tax. That should reduce time and durations of the queue experiences. Otherwise, a paid-holiday for the time it takes to cast the ballot successfully can motivate some positive responses. It is true that the principles risked hanging, incarcerations, poverty, when they signed the Declaration of Independence. Nevertheless, in modern time, voters should not be penalizing with a day's check to cast their vote as a legal right.

Natural Gas

Multiple gas pipelines blow-ups in urban neighborhoods are also bad enough. The none-sleeved unmonitored fracture-gas drillings are all over the southwest and western rangelands, and the consequential underground water pollution damages are still unstudied. They are the uncovered- calamities.

Wildcat Fracking has speeded up over the last decade, as technological injections like multiple directional drillings, and demand for cheaper energy has super charged the political economy for energy and energy alternative solutions. The unregulated wildcat fracking will disposes millions of acreages of land-multiuse values, by pollutions of aquifers and some subterranean water table streams.

In view of exponential growth of fracking and controversies associated with the technological injection of fracking globally. It has become an essential argument for sustainability to study all aspect of new technological injection, not to terminate it, but to learn all safeties associated with its applications, in line with sustainability conditions in service of humanity.[3]

Figure 9 shows fracking and pre-fracking requirements for United States.

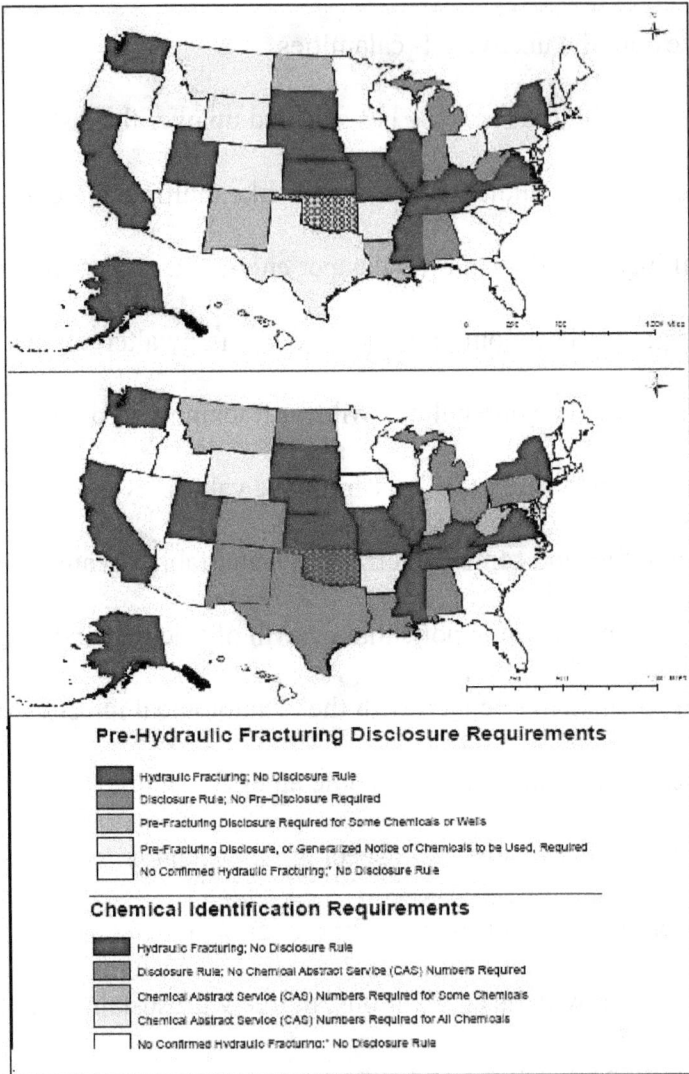

Figure 9, All states fracking requirements.

Individual ranches some under educated, some isolated and under-represented in Western and heartland American States and some victim of cowboys' culture of heroic individualism and austere self-reliance will be bamboozled and hornswoggled by the worst of cooperate city slickers in full operation mode. They sign contracts of their land uses, and unconditional long-term fracking and resource and subterranean land usage contracts, while the economic down cycle has them in their temporal vulnerable economic condition. It will not social welfare function well, to abandon them to skilled and quick operators without any formal legal structures.

Coal

The coalmine cave-ins all with multiple human sacrifices are common to pages of history. Nevertheless, when they all bunch up and occur in the time span of one year that is economically too significant to go nationally and internationally unattended. Open mines, mountain top coalmine have also accumulative-pollutions associated with their production. Toxic run offs and air pollutions

implications follow them with good reasons. The negative

externalities are ignored, or they are sometime bribed away.

Never the less, somebody will pay the full price, dead

miners after the cave-ins, children and elderly with Asthma

and pulmonary condition in the next counties will never be

compensated enough to relocate.

Therefore, the concept of cooperate taxation at

national or federal levels can no longer be ignored. It is

unconceivable that a third major war is unfolded, yet the

so-called representative of the nation so concerned with the

deficits have chosen silence rather cooperate-minimum-

taxation to accommodate their magnanimous gestures of

America and Western humanitarian exemptionalisms.

Since, they have failed at the historicists-supremacists'

exceptionalism.

The long lists of military police-actions were adding

costs to these calamity mixes to secure other nations'

resources on the cheap (with or without United Nation

approvals, or conspiratorial obfuscations of their prolonged human rights stewardship).

These wealth-confiscatory plans and the upward redistributions are not just uber-calamities on Islamic and developing, and third-world nations, the plans are exposed in the cooperate game plans unfolding in the United States against the middle class America, professors, teachers, police, firefighters, and all workers at certain States' governors' rightist (illegal) strategies.

This is to add to the cooperate profits, executive bonuses, and to remedy their cooperate and national balance-sheet melodies, it is time to impose national and an international cooperate minimum tax to pay for the overt violations of human rights and imposed manufactured disasters in negative externalities, at least at the very high level of revenues.

Nuclear

The assertions of safe nuclear energy are false scientific notions. It is a bad scientific hypothesis that the carbon free nuclear energy is safe, or less costly, or

pollution free. None of those hypotheses is true. There are

far too many examples of natural and manufactured unsafe

practices and accidents, which make the safety assertions

false. The relative merits of all energy resources have to be

seriously studied. However, these research and studies

cannot be relegated to the politically handpicked science

salespeople.

Intermittently, the concept of technologic

innovation is used to make assertions of certainty about the

safety of nuclear facilities, in presence of dynamic risks,

entropy and skill degeneration in human capital.[4] That is

even before greed, penny-pinching hubris, pride, and

ignorance intrudes as human failings in the elemental

scientific equations. Demanding perfections beyond

human biology in certain technology cases is simply

unaccepted. That 0.001 percent out of the 100 percent

failing brings 200 percent costs, human mortality, dynamic

genetic damages, and regionally abandonment of resources

in perpetuity. Some chains of unwise notions and decision points are as follows.

Putting 6 potential Chernobyl(s) next to each other, in the radiation fallout kill-zones of each other is one of those bad initial ideas. If one fails other are forced to fail, or workers are in ultimate danger for their lives and so are everyone else up to 50 square miles around them, more or less. Combination of six perpetual spent fuel rod-pool ponds on the top of each structure in an earthquake zone is another bad notion. The concept of six secondary backup water pump generators in tsunami regions at lower than 35 feet above sea level is yet another bad idea, in tsunami high-risk zones. The concept of six sets of backup batteries, with limited life to support spent fuel, in water level, which must stay at full level, and in dynamic risk situation is yet another false notion of safety. Therefore, all great technological injections bunched or placed wrongly yield great buzzwords, but the same bad results.

Now, the other mistaken notions endogenous to the nuclear plants carry no less fundamental risks with them. Brittle metal syndromes associated with radiation are prominent dynamic risk issues, which require dynamic attention. There goes the handy assurance of double chamber concept, pushed and reposed in these six nuclear plants in Japan, the other 25 nuclear plants in USA. Beyond certain date, certain points of entropy the metal chambers no longer hold explosions internally. The following risk probability calculation is another fundamental first step before (re) licensing or construction. It may be an underestimation of the risk percentages.

$$\sum Pi \times (I) = Pi1I1+Pi2I2+... \approx 30 \times (0.01) \approx 30 \text{ Percent}$$

Therefore, the improbable hypothetical chances of risk-event at 1 percent theoretical failure out of 100, and the 99 percent chances of success change, per one event, per each factory changes to higher degree of certainty of negative event in summations. When we add the entire events that must go right for success to occur every time, those improbable chances of failure together are not 0.01

percent, they have mushroomed into a 30 percent very

likely, failure, and70 percent chance success out of 100

events. That is a clausal disastrous outcome (catastrophes),

every 30 years out of 100 years of operation, to study it or

state it differently. The size and ultimate costs of nuclear

disasters make them untenable, too expensive per kilowatt

of electricity, or chunks of earth and human lives we have

to give up to attain such an energy dream state! Is 50

square mile chunks of New York, New Jersey, or Santa

Barbara, St Louise Obispo, and Los Angeles are really can

be abandoned? Some nation's states have decided to

answer that question, because one accident will cover all of

those nations for time immemorial.

The cement outer shell cannot withstand explosions

common to accidental rod exposures, and abnormal

earthquake, tsunami, tornadoes. You put the metal

chamber and the spent rod chambers under the same roof.

You get one shell protection per each, and double the

dynamic risk of accidental harm for both. When you have all or nothing dynamic risks, you need to design for the eventuality, which insures all or nothing events. That is, if there is a chance of 9.0 earthquakes, you use that. If there is a chance of tornado and an earthquake, and loss of electricity, you use that. If you built, in tsunami region, which has 9.0 earthquakes and loss of electricity, you use those.[5] That is, you have to build to absolute maximum of all risks, because the cost is so absolute.

While all energy resources have associated risks, the prepaid government and industry protections of human lives around the nuclear disaster plums are never supposed to have happened! Yet, here we are the day after, Japanese second mass radiation, Utah and New York down winders of post 1950s, after the Chernobyl, and Three Mile Island, and so forth. We need to talk dynamic risks and cooperation's' responsibilities and liabilities.

Post Radiation

After billions of taxpayer spent on these wonderful safety mechanisms, we will ask for the respected governments, IAEA, to give quick measurements associated around the point of disasters, the measurement, and direction of the plums, and elemental makeup of the radiation plums.

Therefore, nationally, and internationally, populations devise their own strategy to deal with the incoming plum-disasters. However, these simple actions have presented many issues for certain governmental organizations in the delivery of the bad news in timely manner. Transparencies are not welcomed ubiquitous universal virtuous, yet.

Technological Injections Stratagems

There are logical points for changes to alternative energy sources in such a way to reduce CO_2 emissions, among other pollutants.

However, this graph gives us a chance for analysis of actual and projected data for the growth rate carbon dioxide (CO2) related to different energy sources used globally.

Data is about the global energy-related carbon dioxide emissions by fuel types [(billion metric tons), for coal, liquids, and natural gas] for periods 1990-2035. They include projected data.[67] The calculations are about the Growth rates (in natural logarithms) of energy-sources of CO2 times (100).

Figure 10 shows the data for growth rate of global CO2 emissions per energy source. In the actual data coal is the out layer data, while liquids and natural gas remain within a level.

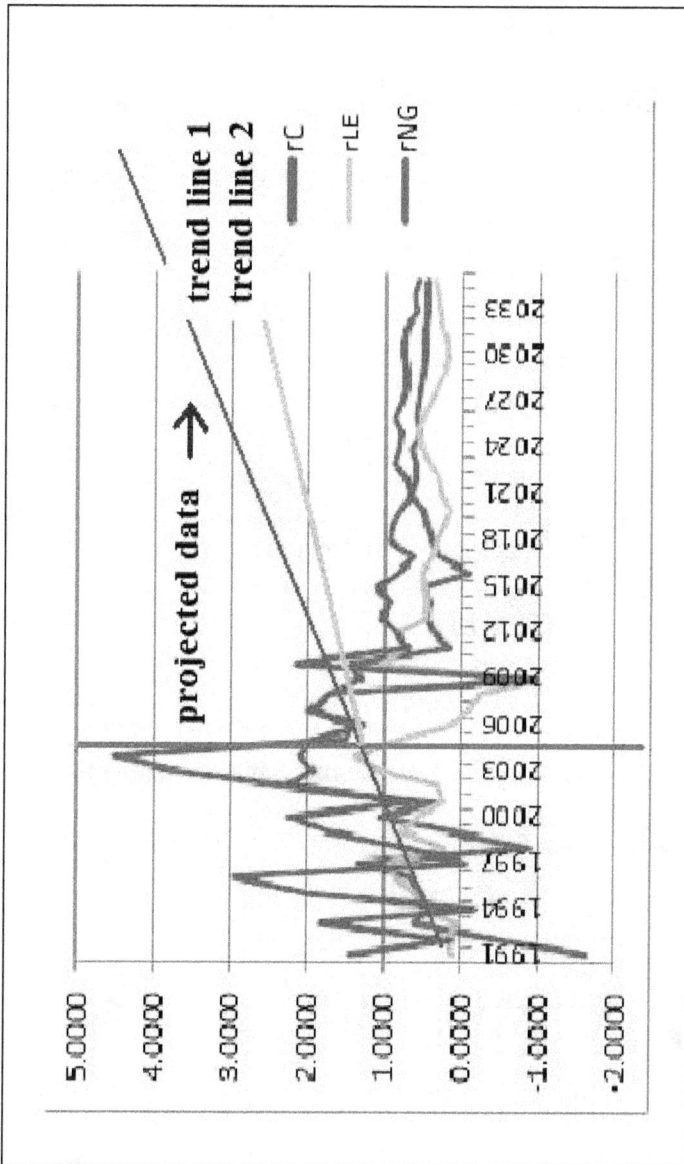

Figure 10, Global CO2 emissions growth rate of energy-sources.

There are better logical grounds behind the

stratagem for change to alternative energy sources. Aside

from the war wariness, financial bankruptcies, or war deficit risks, there are many reasons why there are constructive momentums for change to multiple-energy-source strategies as national security policies, globally.

Some of these reasons are as follows,

- Because supply shortage condition alerts,

- Increase in technological supply of functional alternatives.

- The risks implications of carbon based energy supplies for the climatic extremism.

Trend lines 2 or even a more hopeful trend that shows constancy will require massive investment in alternatives.

Teconometrics of Dynamic Financial Risk and Rewards

Mathematical economics (teconometric) treatment of dynamic financial risks and rewards of all capital investments is as follows.

Limited assumptive conditions, as the cyclical nature of markets are present. Price risk, government risk, currency valuation risks, supply and demand synchronization risks, and climatic and production hazards are assumed exogenously estimated, or given. Sheppard lemmas and Dr. Bahman Fakhraie's sustainability conditions also apply.

$$R = \text{Max } [(Xi) + (x3)] - \text{Min } [(x2)] \geq 0$$

X- av. est. of capital investment returns

$$R > 0,$$

$$1 > P (xi\text{-}n) > 0$$

$$P (Lj1\text{-}jn) > 0$$

R- Revenue

L- Risk loss

P- Probabilities

Figure 11 shows us the comparatives of teconometrics of dynamic financial risks and rewards for global capital investments.

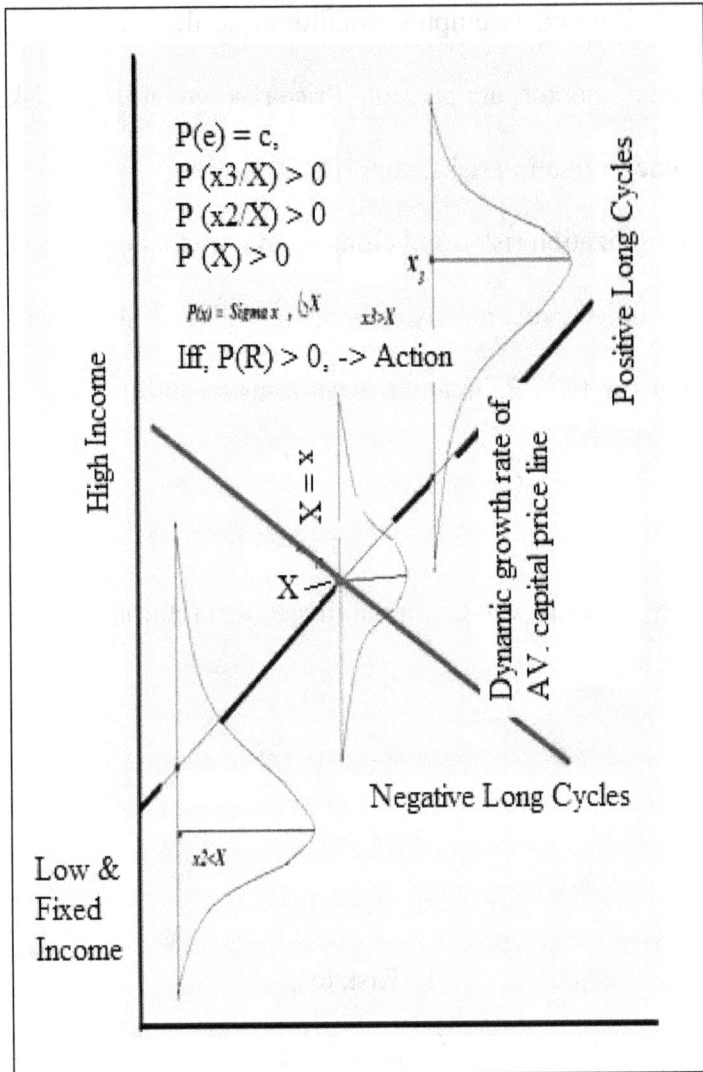

Figure 11, Teconometrics of dynamic financial risk-reward of global capital investments.

Therefore, the strategy of change to alternative

sources is a reasonable target for technological injections, not sequestrations, or deficits-political brawls to a no growth draw. Inaction is hardly a good action strategy.

The comparative juxtaposing of national infrastructural needs and some industrial and related disasters and events should have constructed the proper prospective for higher comprehension. Given the magnitudes of supply requirements that all great society, but especially one as our modern great society require for just maintenance, let along for encouraging a path for growth orientation that keeps with natural global rate of growth.

Figure 12 shows the urgency and need for investment in the infrastructure as national investments. That involves private and governmental infrastructure investments, beyond subsidizations of obsolete technologies in place.

The Urgent Matter of
Infrastructure National Investment Plans

Federal regulators say old wrought- and cast-iron natural-gas pipelines are the most at risk of corrosion, cracking and catastrophic rupturing.

STATES WITH THE MOST: Michigan ranks among the top 10 states for miles of old wrought- and cast-iron pipelines that federal regulators say are high risk and should be replaced.

OPERATORS WITH THE MOST: DTE Gas was the utility with the second most old wrought- and cast-iron pipelines in the U.S. in 2011.

#	State	Miles		#	Operator	Miles
1	New Jersey	5,138		1	Public Service Electric & Gas (N.J.)	4,202
2	New York	4,541		2	DTE Gas	2,499
3	Massachusetts	3,901		3	Boston Gas	2,123
4	Pennsylvania	3,260		4	KeySpan Energy Delivery - New York	1,674
5	Michigan	3,153		5	Peoples Gas Light & Coke, Chicago	1,544
6	Illinois	1,832		6	Philadelphia Gas Works	1,542
7	Connecticut	1,509		7	Niagara Mohawk Power	1,486
8	Alabama	1,452		8	Baltimore Gas & Electric	1,333
9	Maryland	1,422		9	Consolidated Edison of New York	1,305
10	Missouri	1,180 As of March 4		10	Alabama Gas	926

SOURCES: U.S. Dept. of Transportation, Pipeline and Hazardous Materials Safety Administration, Pipeline Safety Trust, City of Royal Oak and Royal Oak Historical Society Museum

Older Local Waterpipes, As Replaced Infrastructures.

Dr. Bahman Fakhraie. © 12.13.2012

Figure 12, Urgent infrastructure national investments

While the figure 12 presents the synchronous disharmony of the material entropy, technological obsolescence, and infrastructure investments of the industrial worlds, compelling arguments exists for repetitious usage of inappropriate structural martial after massive earthquakes in most of the global hot spots.[8] The use of mud, hey and gypsum, and unenforced concrete, breaks and iron-beams are noted in most earthquake zones, where even industrial building crashes.

National investments in such infrastructures are prudent because they will prevent some other future disasters and industrial disasters, if instead of the usual expo facto blame games, nation's politicians use sound planning and infrastructures investments, and pass helpful constructive laws, which will enhance national security and create jobs.

Notes

[1] Author calculations from Sources: AEO2013 National Energy Modeling System, run REF2013.D102312A; and AEO2012 National Energy Modeling System, run REF2012.D020112C, http://www.eia.gov/analysis/projection-data.cfm#annualproj

[2] Dr. Bahman Fakhraie, *Teconomics: the microeconomic analysis*, Utah, FERDAT publishing 2012, and, https://www.createspace.com/4196760, Books web page at, http://bahfecon.wix.com/bahfecon

[3] M. MCfeeley, *State Hydraulic Fracking Disclosure Rules and Enforcements: A Comparison*, Natural Resource Defense Council, NRDC Issue Brief, July 2012, IB:12-06-A, http://www.nrdc.org/energy/files/Fracking-Disclosure-IB.pdf

[4] Zoltan Horvath, Tribute_to_Chernobyl_disaster_-_Sleeping_Sun, History Channel

[5] RT, 2011_Sendai_earthquake_and_tsunami_No.3

[6] Natural logarithms of growth rates and trend lines 1, 2 are estimations by author.
"Energy-related carbon dioxide emissions,"
U.S. Energy Information Administration
International Energy Outlook2011
DOE/EIA-0484(2011), July 2011
http://www.eia.gov/forecasts/ieo/emissions.cfm

[8] Iranian, Chinese earthquakes, and Bangladesh 2013.

CHPTER 3

General and Specific of Disasters

There are three periods associated with all disasters, before, during, and after the disaster- periods. However, they are not always distinct, or long enough to service detail planners, or allow for long perpetration-periods. Nevertheless, an individual who train and prepare and approaches these disasters with plans tend to do much better and survive them. Large plans by cooperation of all governmental entities, at local, state, and federal level unit behind improve the chances of success to notable levels, indeed.

The action lists in cases of disasters are limited but they are not null.

- All disasters: preplanned list for the disasters, remain situation alert but not neurotic, at all times.

Keep an emergency checklists, start the lists with all human lives are precious, preserve yours to preserve all others around you, that almost all tangibles are replaceable.

• All disasters: preplanned list, have documents, some money, credit cards are better, and information for quick access. Have 48-74 hours of water and food and none perishable cans of food items on a recycle bases, use old one (check their expiration dates) and replace them with new items. Keep first aid items, flashlights, batteries, radio. Keep outdoor supplies as for camping, or gas grills supplied for events. Repair and maintain, gasoline-chainsaws, and electrical saws, and keep supply of gasoline cans for different operations.

• All disasters: during the disaster, clean and safe foods and supplies are a priority, before the disaster, and short term after the disasters. Keeping groceries open and making arranging to transfer food and supply to locations, logistically. Supermarkets, warehouses, have taken these logistical arrangements, much more seriously recently.

• All disasters: opening triage medical centers immediately after the location is safe and longer term logistic arrangements for health cares are top priority. Make access to physical and mental health as priorities.

• All disasters, expedited safety and policing, firepersons, and government people, federal, state, localities, and businesses people, cooperate services, in insurance, clean ups, coastguards and FEMA, property protections, and debris removals, and Red Cross services are all priority.

• All disasters: immediate aftereffects of the disaster-check on family, neighbors, infirmed, and older people around your area are important service. Do the good head counts? Who is there? Moreover, are they all OK, or do they need services?

• Nuke disasters: Governments, The 100-kilometer diameter of relocation of pregnant women, children, up to teens are most commonly practiced, and as response to Chernobyl data. Clean water and food are also helpful to comfort and survival.

• Nuke disasters: Governments, use of Potassium Iodine pills and drops ONLY for those in immediate exposure range, of dangerous plums, for the limited time 24, 44 hours danger time, until you are out of the exposure zone, and as prescribed by a physician. One a day, multivitamins with full elements often have the same (Potassium Iodine) in them. They are a better choice, with less sickness side effects. Those concerned and not in danger of high radiation, the use of such multivitamins are advised. Hording the pills are useless and counterproductive.

• Nuke disasters: individuals, staying indoors, out of rain or snow in radiation-zones, or washing the skin with clean water, after exposure are also common safe practices. Avoid contaminated milk, contaminated fresh fruits, food, and vegetables, as long as, they are contaminated.

• Nuke disasters: the existence and presentation of actual radiation-measurements are the best solution to ignorance-induced panics. Currently EPA internet link gives a data streams.

• Petrochemicals, spillage and fire fumes disasters, moving away from the zone, at the authorities requests are recommended. Short term, staying indoor, shutting windows, with air conditioning may work for a few hours.

• Major regional and metropolitan earthquakes are common. The risks are explained as one in the lifetime events. I have been in at least three 7-points earthquakes, and one over the 8-point earthquake. It helps if there are water bottles for 24-48 hours, groceries, wholesales warehouses may open their doors for resupply, water bottles, and some fresh food, and nonperishables that will help.

• Major regional and metropolitan earthquakes, traveling is extremely dangerous, especially immediately after the disaster. Because of cave-ins, bridge damages, in the roads. If you have to be on the road, make certain of the road ahead by traveling very slow.

• Major regional and metropolitan earthquakes, keep some documentation, identifications, and credit cards handy for easy pickup on your way out of the buildings to safety. There are recommendations to video record your apartment-contents for insurance reclaim purposes. Nevertheless, remember all lives are irreplaceable, and all tangibles are almost replaceable. So, preserve lives.

• Major regional and metropolitan earthquakes, it helps if you can shut the natural gas at the gas meter, and turn the water heater and the heater back on after the dangers has passed.

• All disasters: Deal with extreme uncertainty and doubts about, food, health, and healthcare, shorter, and longer-term aftereffects.

• All disasters: Streamline emergency follow-ups for government and cooperate services, insurance, documentation duplications, and reports.

From the futile cold war days, of duck and cover to modern days, science and technologies have advanced by leaps and bounces, yet the human bad notions, even of experts, and science sale people do not always keep up. A big city in Japan did an exercise disaster-test, in one of the most trained population for major disaster urban center. They attempted total population evacuation-tests of a major metropolitan, to see how quickly, on emergency declared bases, they can evacuate a major city. The results of study concluded that it took so long that it shocked and awed all observers.

This is noteworthy, because United States military and some scientists for a long time during the cold war were under the same notions. The authorities will evacuate big metropolitans' cities, before a 20-minute long nuclear missile attack. Well, after a few years in 1970s, some foreign scholar reminded them, if the afternoon rush hour commutes takes from 2-7 PM, without emergency. How can emergency evacuations take less than 20 minutes? The responses were a quiet burial of the mistaken notions, and initially a doggy global peace movement for nuclear disarmaments.

Therefore, it took 3 days and some hours, to evacuate the large Japanese metropolitan city on an emergency bases using city transportations, metro trains, and trains. It will be surprising, if another nation beat them in this record. That will explain the scientific inflection-point beyond the antinuke movements globally, but, in Japan and Germany specifically.

Hence, neither the duck and cover, nor the evacuations of metropolitans' modular production disaster plans met the Sheppard lemma, or the sustainability conditions. Therefore, caveat emptor is called for, as much as Adam Smiths' other thoughts are underutilized, even among the so-called fake political opportunistic conservatives.

In this scientific paper, *"A Treatise on Energy Resource Production and dynamic Risks Of 2011 Quatsunuk (QUA'SU'NUK) Disaster and Index"* for the 2011 Japanese (Toceh) disaster, a new name [2011 Quatsunuk (QUA'SU'NUK) disaster] was suggested.[1] A new Quatsunuk risk index was also presented for high level 6/7 nuclear disaster (86 percent), plus 8.9 out of 9 level earthquakes (99 percent) and 35 feet tall tsunami out 35 feet tall tsunami (100 percent) risk were also presented as a good start for the Quatsunuk index. The arithmetic-average index calculation will yield a 95 percent QI (Quatsunuk Index). If the nuclear disaster proceeds to level seven, the Quatsunuk Index will become a 99.6 percent QI disaster index.

In either case, this is the biggest measured known combination disaster to scientific community. For future epistemological ease, Quatornuk index (QtI) will be used, where tornado is in the disaster mix, and (QsI) can represent (Quatsunuk Index) for risk and damage presentations.

Modular Production Processes

The nature of production functions and processes have changed.[2] However, the need and requirement for their continuation have grown exponentially. The enumeration of risks and equivocation of GNP growth and CO_2 are not arguments for termination of production processes.

The nine billion human population growths dictates a balanced path to the needed continuation of such modular production arrangements, and associated international trade for the future of human history.

In this case A, there is positive chance for dynamic risk, $\frac{dr}{dt} > 0$

1-In cases of service, good, or investment qi, there is a risk-chance qi may be good or bad, (sd = δ, σ, $\pm$).

2-π = f (p, q, cq, rq) = f (+bq, - cq, - rq) = (p. q - c)

The positive possibilities are not clairvoyant, however, sigma control and improvement in production, and standard deviation of risk in investment and major disasters can establish some benchmarks out of the darkness associated with repeated disasters, Pq, and q = f(σ, $\pm$), and rq or I = f(δ, $\pm$). How we enumerate associated costs and negative externalities will prepare the system for the next onslaught.

Governments and Policy

Multinational energy wars with millions dead in a Middle East Islamic nations of Iraq, Afghanistan, and Pakistan, a chain of induced and self-induced revolutions in Islamic nations with victim fatalities are bad enough, to continue on the same misguided roads to further calamities is very unwise, aside from being extremely costly in the longer run. Criminal conspiracies and perverted political collusions to undermine human development and their

economic growth, under arranged counter democracies

have met some technological stall.

Those dishonest and unfortunate pervasive-resource

striping arrangements of the past years in the presence of

technological transparencies have come to a dead stop.

Certain business and cooperation-arrangements to engage

in production under unsafe and globally fatal conditions,

under market-imperfections, for patriotic machinations,

militaristic straw men menageries of excuses will meet the

same shameful, fatal, and disastrous outcome as seen.

The next graph presents the reduction in production

and consumptions and equilibrium level Et1, to the post

disaster level Epd. The degrees of recovery will depends

on both private free market, and governmental (national

and international assists and aid), producers, individuals in

the disaster zone will receive, in order, to make them whole

at E1, at t2 time. That is sometime in the future, which will

be a lost. In the other hand, private markets can help with

recovery loans, and technological assists, which will advance the individuals to their higher social welfare position at E2 equilibriums, unless imperfect (political and economic) markets conditions persists. Figure 13 shows these post disaster social welfare utility positions.

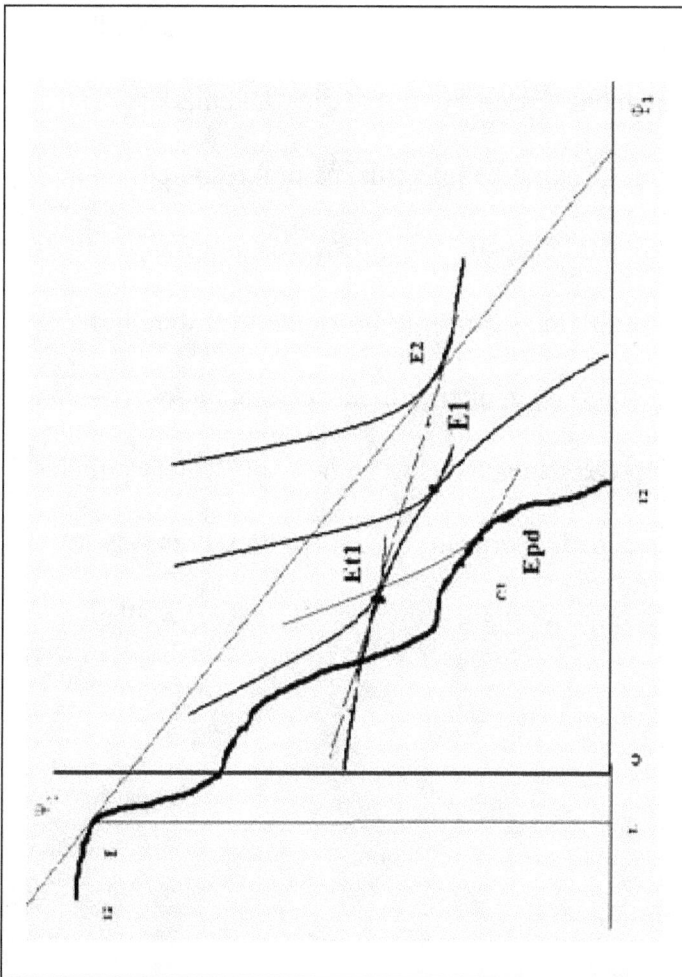

Figure 13, Post disaster production adjustments.

There are going to be naked aggression against the technological inevitability and the evolution of practices and policies, already political well on their ways. Most of theological counter culture wars to hold back progress will again fail. On the other hand, misapplications of technological innovation to spy, monitor, and holdback employment, dissent living salaries, and marketable resources, capital, labor skills, natural resources, in an equitable redistribution of global wealth creation (different from wealth miss concentrations, harmful wealth, concentrations) in international suitable trades will also come to the same dead-end , other confiscatory policies have reached. Humanity is not able to kill its way to halting human progress toward humanities' wellbeing.

Notes

[1] Dr. Bahman Fakhraie, *"A Treatise on Energy Resource Production and dynamic Risks Of 2011 Quatsunuk (QUA'SU'NUK) Disaster and Index,"* scientific paper, 10,28,2011,

[2] Dr. Bahman Fakhraie, *Teconomics: the microeconomic analysis*, Utah, FERDAT publishing 2012, and, https://www.createspace.com/4196760, Books web page at, http://bahfecon.wix.com/bahfecon
Dr. Bahman Fakhraie, *Technological injection, dynamic new capital measurements, and Production Theory in Economics*, (Michigan: ProQuest LLC, 2010) and, https://order.proquest.com/OA_HTML/pqdtibeCCtpItmDspRte.jsp
Books web page at, http://bahfecon.wix.com/bahfecon

CHPTER 4

Summary and Some Conclusions

President John F. Kennedy stated, "Problems are people made and people can fix them." It is true now, that people can make their problems much bigger by technological injections. Nevertheless, the technological injections can also assist to undo a lot of the harm, as well. If we allow scientific evaluations and established assumptive truth guideposts get to the people. Greed will not countermand or replace the facts lightly. There is a need for further detail studies of embryonic non-carbon energy resources, and transformation stages and plans of action. The size of such a project and its colossal significances as a technological leap scribes it as a government funded initial mega projects. However, it will be a mistake to assume, their numerous potential profitable and commercial offshoots can be contained. The positive

externalities associated with such technological injections

will permeate and benefit the free markets globally.

Private and Governmental Constructional-ism

Once again the economic science discoveries guides

towards post-Schumpeterian macroeconomic private and

governmental reparative constructional-ism.

Another wartime president Dwight D. Eisenhower's

war experience alerted us to another pitfall of history.

Nations can waste time and efforts on political and deficit

polemics, while wasting trillions harming other human

beings around the globe, and still end up in the darkness of

one kind or another, and economically broke, and still

without prioritized viable solutions. After all, they still

need to spend fortunes and help others to rebuilt back to the

same place of history, where they lost their sanity in the

first place, or where nature took them.

The interstate highways infrastructure-projects part

of government-private infrastructure-plans of President

Dwight D. Eisenhower, a republican war hero propelled USA economy into present millennium. American voters should show little patience for those whom wear down his political party label with small-minded narrowness of visions and plans, and miserliness in their futuristic views of American national purpose, as their own limited visions.

Governments as almost all manufactured tools, systems, and services have appropriate functionalities. There are skills, educations, retraining, job creations, and upgrades associated with all technological injections, which will modify their negative externalities. There are luddites whom are displaced by new technologies. That is why governmental tax and transfers were designed to assists. It is the height irony, if they are the same group, which scream the most about governmental expansionism.

Figure 14, shows the influences of technological injections in a fully functional economy.

Figure 15 shows dynamic circular flow in the economy and important of governmental presence.

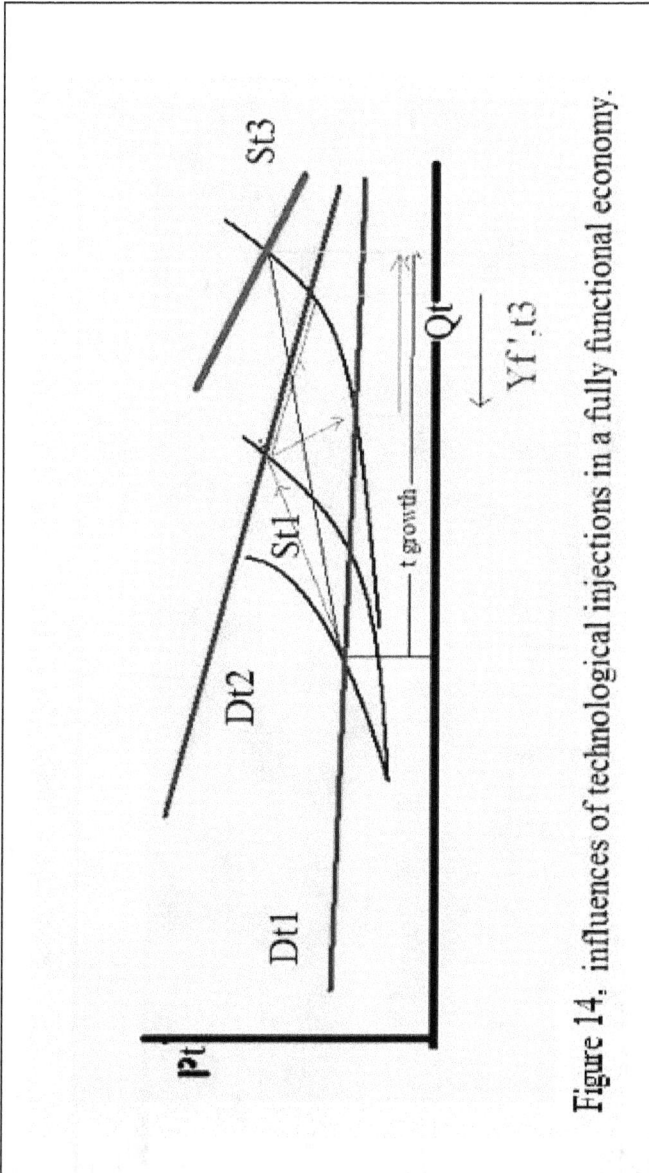

Figure 14, influences of technological injections in a fully functional economy.

Figure 14, Technological injections in functional economics.

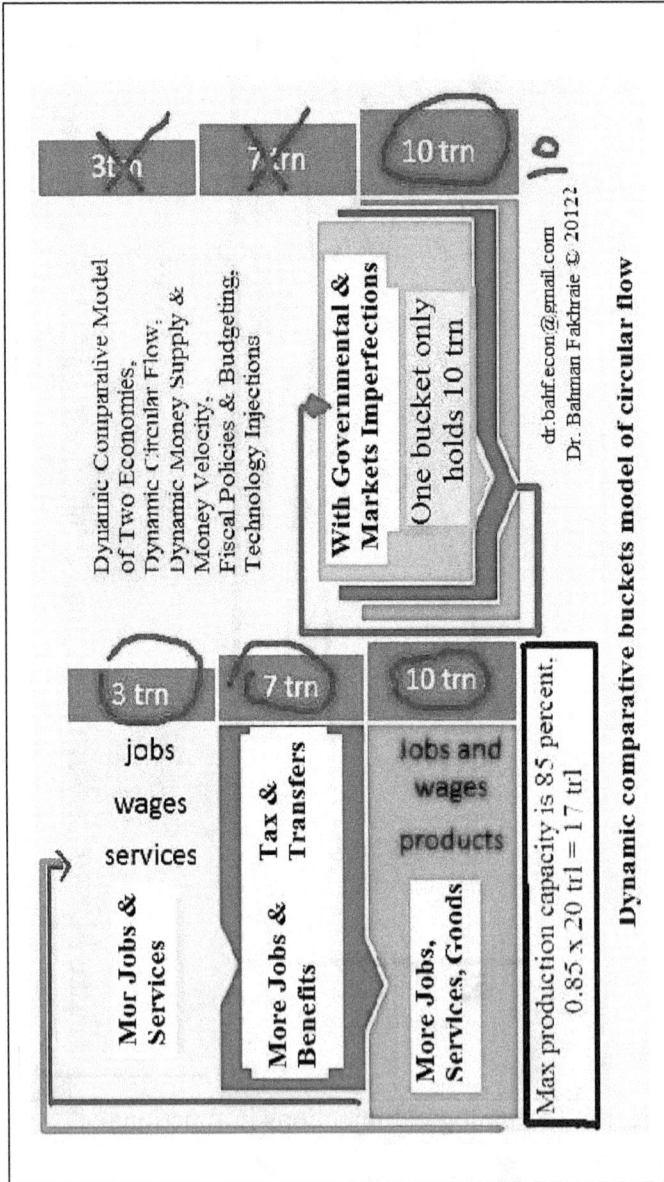

Figure 15, Dynamic circular flow in the economy.

The rest of population can agree with them about the intrusions, the taxation without representations, militarizing of domestic economy, legislative process, judiciary, commerce, and all other malfeasances. Nevertheless, on retraining and job creations, investment in infrastructure, health, and public safety issues, investment in research and scientific discoveries, and on educational opportunities, they are mistaken.

People do understand very well, that nature will humble humanity at every bad turn, that is why humanity comes together and combines their energy, resources, and funds (tax and transfers by governments, decent livable salaries by private businesses) to rebuild.[1] This is called redistributive value theory under economics subjects, yet it makes distinction between constructive wealth creation, and nonproductive wealth concentrations. Even in the modern none Marxists unorthodox economic theories complement to orthodox economics, including the classics.

It is important to realize human economic history evolved from holding gold and it immobility, to using notes (a promissory note) about the amount of gold holdings, to notes as paper money, for the influence of increased velocity of money in expansion of commerce. Both the Republicans (Presidents Lincoln civil war greenback notes, Nixon abolishing gold standards) and democratic presidents (FDR) upheld these progressions.

People depend on scientific discovery of nature, and allow for governmental combination of reasonable resources and revenues to countermand natural disasters, and people made economic and financials' cyclical calamities, so economists devised methodologies to countermand the cyclical calamities, with some degree of success.

The federal, state, and local governments are all of us, so are the insurance cooperation under private contracts that pay for some rebuilding, warehouses, grocery stores, supplies of goods and services are linked to us by the economy as well. All these businesses require the government services to reestablish their market links and supply routes, and infrastructures, before they can get there close to all of us. Nevertheless, it will take the governments some time to get there, under national and natural emergency conditions.[2] A set of modular production processes starting at thought-action level, and fulfilling all three lemma, including sustainability lemma.[3] Dr. Bahman Fakhraie presented those conditions, in the studies that advanced the new lemma of sustainability.[4] This was as an addition to Sheppard lemmas necessary and sufficiency conditions for new economic orthodoxy of production optimizations.

The Global Dimensions

Natural disasters and catastrophes unite humanity very quickly to face calamitous occasions in unity and solemnity. An admirable humane quality used by humanity too miserly at times. The principle is not new by any means. A thirteen-century Persian poet Saadi of Shiraz

Iran wrote the following poem.[5] There are different interpretations, translations, connotations, and meanings from this old poem in different languages. The different meanings in deep and divine poems are common to traditional Persian poetry, as student of the Farsi language know this well. This is the author translation to English of this famous Persian poem.

"People are each other's limbs and lives,

   In creation they are produced from one jewel (essence)

As events, pain one limb,

   All other limbs feel the discomforts and pain

You that do not empathy for others misfortunes and Pains,

   Make it impossible to count you as part of human race"

   Another reason the poem is famed, it is at the entrance of United Nation building. The following figure shows its Persian Farsi language in cursive Nas'taliq Farsi writing.

Figure 16 shows Sa'adi's poem in Persian calligraphy, which is ascribed at the entrance of United States Building.

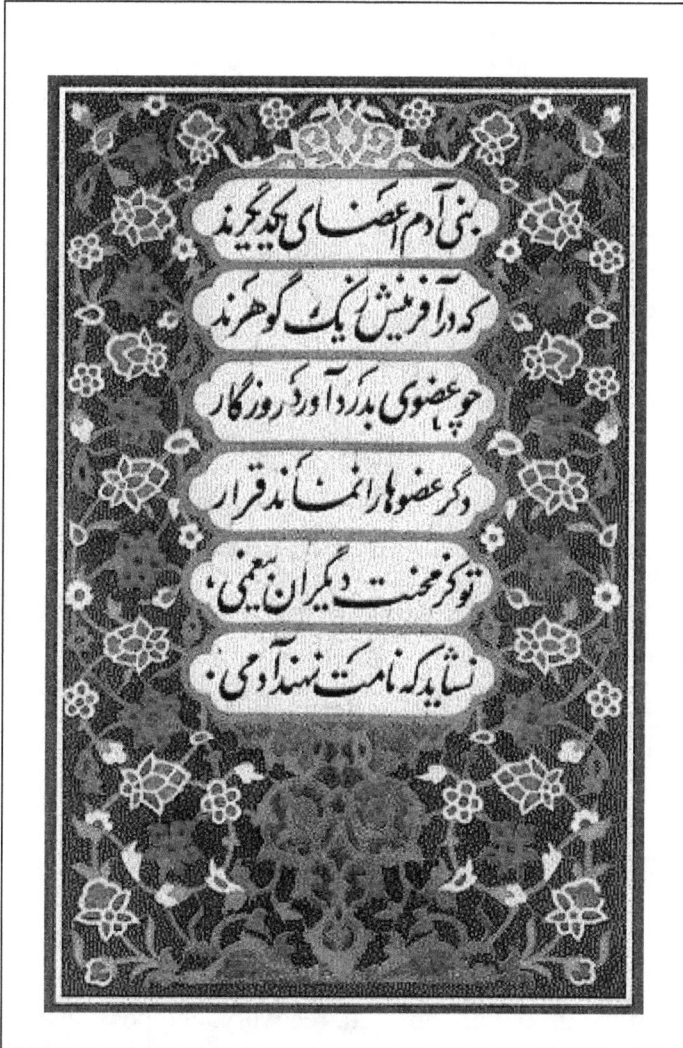

بنی آدم اعضای یکدیگرند
که در آفرینش ز یک گوهرند
چو عضوی بدرد آورد روزگار
دگر عضوها را نماند قرار
تو کز محنت دیگران بی‌غمی
نشاید که نامت نهند آدمی

Figure 16, Sa'adi's Persian poem at United Nation building.

Iran has been the unfortunate host of many natural

disasters in form of earthquakes with fatalities over time.

Nevertheless, It has managed to come back repeatedly, and

sometimes with the help of international friends.

Therefore, the words of Secretary General are particularly

kind to Iranian people.

> "Persian cultural and artistic traditions have enriched
> humanity for centuries – many hundreds and thousands
> of years. Your poetry is unrivalled in its imagery and
> depth of feeling...
> Persian language itself is a poem, so whoever or
> whenever you speak the Persian language, it sounds and
> looks like poems. I agree with that. Your architectural
> wonders have filled people with awe - from the
> explorers of old to travellers today.
> Iranians are rightly proud of these achievements.
> As one of our founding members of the United Nations,
> Iran also has a long history with the United
> Nations. Through the years, we have worked together
> on many issues of concern."[6]

Therefore essential planning, multisource

dependencies for funding and supplies, including uses of

naval task force and military assets all become essential

parts of the gestalt of survival, recovery, and post

Schumpeterian policy of constructional-ism. As the global

dimensions shows increases in frequency and occurrences of disasters, and global catastrophes, a plan of action, which include humanity will become more of a welcome scenarios among all of the squabbles and incessant warfare of modern humanity. Advancement of knowledge encourages a departure of beguiled virginal innocence of the past agendas of hegemonies, misapplications of United Nations' resources to longed induced wars and conspiracies, which will do very little to improve humanities' survival rates, in the age of multiple catastrophes.

Is an observation, that nature does not welcome or is kind to intellectually dispersed competition, when it comes to disastrous ruinations. Nature will not feel the pains of humanities' limbs. Even more reasons humanity needs to unify their resolve and efforts dynamically and to enhance constructionism cooperatively in the face of increased statistics of disasters and catastrophes globally.

Notes

[1] Dr. Bahman Fakhraie, *Analytical Remedies for The Millennial Cascading Economic Declines,* Utah FERDAT Publishing 2012, https://www.createspace.com/4187823, Books web page at, http://bahfecon.wix.com/bahfecon#

[2] Dr. Bahman Fakhraie, *Teconomics: the microeconomic analysis*, Utah, FERDAT publishing 2012, and, https://www.createspace.com/4196760, Books web page at, http://bahfecon.wix.com/bahfeconE

[3] Dr. Bahman Fakhraie, *Technological injection, dynamic new capital measurements, and Production Theory in Economics*, (Michigan: ProQuest LLC, 2010) and, https://order.proquest.com/OA_HTML/pqdtibeCCtpItmDsp Rte.jsp, Dr. Bahman Fakhraie's books web page is at, http://bahfecon.wix.com/bahfeconE

[4] Appendix of this work continues the mathematical presentations.
Teconomics of Verbalism, Utah FERDAT Publishing 2012, and at, https://www.createspace.com/4121720,
Books web page at, http://bahfecon.wix.com/bahfecon

[5] Saadi Shirazi, Sheikh Mosleh al-Din, 1200s-1292, educated locally at Shiraz, later in Baghdad Islamic school, was a traveler of like Marco Polo in Middle East and Africa, but he favored one to one communication in fashion of dervish, Sufis, with common people. He published Bostan, (Orchard), and Golestan, (Rose Garden). http://www.iranchamber.com/literature/saadi/saadi.php

[6] "Tehran, Iran, 30 August 2012 - Secretary-General's remarks at the School of International Relations" http://www.un.org/sg/statements/index.asp?nid=6266

Appendix

A

ARITHMATIC OF OPTIMIZATION

The series of mathematical analysis are presented henceforth, to study utility optimization and demand side, and production optimization and the supply side analysis. The traditional mathematical evaluation for optimization analysis is as follows:

Traditional productions function,

$$o = f(v_i^\varepsilon)$$

Necessary condition, it is also known as the first order condition (FOC):

$$f'(v_i^\varepsilon) > 0$$

Sufficient condition, it is also known as secondary condition (SOC):

$$f''(v_i^\varepsilon) > 0$$

The traditional production function is with mathematical

evaluation for optimization analysis.

Conditions in Case of Growth

$$o = f(v_i^\varepsilon)$$
Necessary condition:
$$f'(v_i^\varepsilon) > 0$$
Sufficient condition:
$$f''(v_i^\varepsilon) > 0$$

Implication: the function is positive and increasing at the
increasing rate.

Condition in Case of Decay

$$o = f(v_i^\varepsilon)$$
Necessary condition:
$$f'(v_i^\varepsilon) > 0$$
Sufficient condition:
$$f''(v_i^\varepsilon) < 0$$

Implication: the function is positive and increasing at the

decreasing rate.

B

ARITHMATIC OF TIME

Lagrange analysis of n-variables and multi-

constraints Lagrange multipliers

The following is the mathematical analysis of productive

use of available entrepreneurial times, on daily bases.

Individuals have to make decision in time-use management,

between productive choices they have to make in use of the

fixed 24 hours.

They make assumptive choices about their time

uses, for required sleep, and time lot spent on education,

and required work for income, as the following relations

show. However, in post emergency or post disaster time,

other arrangements have to be made to replace the

collateral losses, and the income earning time. Churches

Mosques, Temples, and governmental and none

governmental (NGOs) have had a historical rule.

$$f(X_S X_E X_W) = Q_i$$

$$g(X_S X_E X_W) = 24$$

$$h(X_S X_E X_W) = y_0$$

$$Z = f(X_S X_E X_W) + \mu[(24 - g(X_S X_E X_W)] + \gamma[(y_0 - h(X_S X_E X_W)] = 0$$

$$Qi = f(Xi)$$

$$g(Xi) = 24$$

$$h(Xi) = y_0$$

$$Z\mu = 24 - g(Xi) = 0$$

$$Z\gamma = y_0 - h(Xi) = 0$$

$$Zi = fi - \mu\, gi - \gamma\, hi = 0$$

Arithmetical Proof:

FOC

$$\vartheta Z^1 = 0$$

SOC

$$\vartheta Z^2 < 0$$

All constants are great than zero.

C

MAXIMIZATION OF GAINS, OPTIMIZATION OF VALUE

Budget determines expenditure on costs of production of X, and Y. Y can be designated governmental, or NGO injection in post disaster time, while X, is some local productions still operational, which must be encouraged to keep the level of economic activities high to accommodate the post disaster economy. λ_i is unit costs.

$$B = \lambda_1 X_1 + \lambda_2 Y_2$$

S denotes level of skills, technological injections required. In this example, λ_2 denote the skill level of NGOs, or governmental units presented at the scene of disasters. In some nations, local skill or technologies are not able to present the type of presence, which the post disaster conditions demand. Geographically isolated areas require naval assets, which are not present in most countries to meet the logistic demands of post disaster conditions.

$$S = s_1 X_2 + s_2 Y_2$$

Education and retraining required in use capital productively.

$$E_k = e1\, k_X + e2\, k_Y$$

Production function is,

$$Q = f(X, Y : \dot{B}, \dot{S}, \dot{E})$$

The Lagrange equation is,

$$Z_Q = f(Vi) + \mu i\, [(\dot{B} - \lambda i\, Vi)] + \mu j\, [(\dot{S} - \lambda j\, Vj)] + \mu l\, [(\dot{E}_k - \lambda l\, Vl)]$$

FOC

$$d Z_Q / dt = 0$$

SOC

$$d^2 Z_Q / d^2 t < 0$$

More assumptive conditions are as follows,

All constants > 0

OPTIMIZATION OF VALUE

Optimization of Value Ψ is a function of maximization of profit, minimization of costs, and optimization of social welfare function of Vi.

$$\Psi = f\,(\text{max}\pi,\ \min C,\ \text{opt Ui})$$

$$Ui = f\,(qxi,\ qyi : Cvi : Bi,\ -/+?)$$

And

$$Ui = f\,(+,\ 0,\ -/+?)$$

COMPENSATORY SUSTAINABILITY CONDITIONS

Individual Utility Function

$$Ui = f\,(+,\ 0,\ -/+?)$$

The utility of individual Vi is improved in consumption of Qxi, while the consumption of Qyi stays the same, and the budget line can improve, stay the same, or become negative, thus also influencing the utility function.

Social Welfare Function of Group Utilities

$$\text{Opt SWFvi, SWFvi} = f\,(Ui,\ Uj,\ Ul,\ -/+?)$$

Majority of group utility will be better off at a new position A, therefore, they will agree to the project. The utility of some of group utility will stay the same. They stay either satisfy or join the last groups at position B. The utility of some of group will be worst off, there has to be tradeoffs, as in exchanges, higher remunerations, or tax-transfer strategies to optimize the experiences of all involved for them to agree to move to the better position A. This is another clause added to the arguments for sustainability conditions in all economic system orthodox or unorthodox, holistically.

SUSTIANABILITY CONDITIONS of

DISASTER-AIDS

Let us take a little time to study some of the sustainability conditions of benefits of disaster aids, with respect to social welfare functions for the group utility functions, for all individual in the group.

Two positions are A, and B, such that A is better than B.

$$A > B$$

Therefore, A is preferred to B,

$$\therefore A \ P \ B$$

$$SWFvi = f(Ui, Uj, Ul, -/+?)$$

The benefits sustainability conditions are as follows,

1. Sustainably condition universal criteria, it is if at least one individual utility function improves at A, others feel no change, then →A, A>B.

2. Sustainably condition Kaldorian criteria, this is when those in position A, pay, bribe, exchange with losers for the move to A, since and A P B, then A > B. Or offer to aid at their time of need.

3. Sustainably condition Hicks criteria, those losing from move to A, cannot bribe the groups to keep them from moving to A, therefore APB, →A, A>B.

4. Sustainably condition Scitovsky criteria, if the gainers can compensate (promise to help later) some of the losers. Moreover, some of the losers cannot bribe any of the winners from the

move to A. The move will take place. Such

that A P B, hence, →A, A>B.

5. Sustainably condition Dr. Bahman Fakhraie

criteria, or the dynamic sustainably universal

criteria, (NGOs or Gs assist production to

operational levels) that technological injection

will improve all utilities positions

proportionally, such that [∀ Ui at A] P B, then

→A, A>B. For some future post-functionality

arrangements, (as tax and transfers).

There are more studies under welfare economic

theory, and potential economic dissertational works.

D

RATES OF GROWTH AND DECAY

$$r^* = (Xt2\text{-}Xt1)/Xt1$$

$$r = \frac{\ln Xt2 - \ln Xt1}{\ln Xt1}$$

$$Ey, ms = \frac{\dfrac{\delta y}{\delta t}}{\dfrac{\delta ms}{\delta t}}$$

FERDAT, AND LEGALS

FAKHRAIE EDUCATION RESEARCH
DEVELOPMENT AND TRUST

FERDT

FERDAT

Fakhraie Education Research
Development And Trust

Organization, Fakhraie Bahman

Grant proposal by

Bahman Fakhraie, PhD, ©, ®, ™

Budgetary, Managerial Directive and Legal Clauses

Legal letter and charges for unauthorized information use.

Section 1, Mission Statement

Section 2, Budgetary Clause

Section 3, Managerial Clause

Section 4, Legal Clause

Section 5, Organization Chart

Section 6, Data Management

Section 7, Vita

Legal letter is in cases of e-fraud, misuse of firms' information, or harm under sec. G. All linked and relaying entities, institutions, governmental agencies are accountable no exception.

Dear Sirs, persons, INC., (illegal use of firms' info), your current bill is **$10,000, ≤Ceiling $60 mils.** On behalf of Dr. Bahman Fakhraie, FERDAT, and the firm, family trusts.

These are the current charges related to illegal e-fraud and all harmful misuse of firms information, after the first time we asked you to terminate all contract and remove our name, firms' name etc. on this date.

1-Illegal use, the unapproved initial use of firm's name, contact name, duns #, any and all information supplied to USA government for the sole purpose of firm doing business with USA government, grant.gov, and NSF, Universities, other governmental and tribal entities, and all grantors. Minimum charges are $10,000.00, up to the ceiling of $ 60 million.

2-Use of firm contacts, names, address phone #, TIN/#s, and any information related to that.

Minimum charges are $10,000.00, up to the ceiling of $60 mils.

3- Continuation after repeated formal request to terminate illegal actions, sending junk e-mail, junk mail, forcing this firm to send back, by mail and certified mail. Legal costs related to that.

Minimum charges are $10,000.00 per each event, up to the ceiling of $60 mils.

4- Interference with firm soul business, time sensitive writing of government and private grants, at grant.gov, and NSF, universities, private foundations, NGOs and all grantors, with 4.5 million dollars floor. The Minimum charges are 4.5 million dollars, up to the ceiling of $60 mils.

4- Absolute or required non-disclosures by the firm or Dr. Fakhraie will require pre-paid plan with minimum flooring pay. Flex-legalism targeted at the firm, or Dr. B. Fakhraie will be due cause for nonpaid termination by the same, and harm-terms of Sec G of private contracts.

6- Punitive damages are up to 60 million dollars, if you do not stop, after the first notification.

Please make an effort to pay your bill. There will be additional charges added and sent to you for each time you try to contact the same entities, after the same dated termination notices.

Dr. BF, this is part of the legal notices under sec G, or formal termination notice dates.

Section 1, Mission Statement

To formulate optimum functionality in management teams, in order to make profitable and beneficial contributions, and create opportunity for merited advancements, **to advance research** in my fields of study. Utilizing rare combination of theory and empiricism, cultivated in multi-cultural inclusive settings in academic and private enterprises, I enhance the management in productivity and completion of productive projects, or suggest corrective recursive evaluation and total quality control improvements.

Experiences, among fields of International Trade, Managerial, and Production theories in Economics, Finance, Personal Finance management, and Businesses, which modern business and academic institutions find very expensive to employ and too costly not to employ. My extensive backgrounds help me advance research in my fields of study.

Section 2, Budgetary Clause

* All initial dedicated deposits (inflows) will be transferred to a FREDT, or FREDAT business account.
* No allotments or contracts activity will take place, until full transfer of funds has taken place per contracts. The cash-accounting method is used for tax purposes.

* Alteration to submit to requirements and regulations will fully transfer all risks, fees, and penalties, and legal responsibilities, to the source of the requirements.

* There will be a minimum product list, or writing, or proposal-output for further research, which will be presented and agreed to at the initial phases.
* This will be the only output legally required at the end of contract, or at the end of grant periods.
* There is a managerial flow chart enclosed to enhance comprehensions of cash flow, read after the legal notes section.

Section 3, Managerial Clause
* Dr. Bahman Fakhraie will act as agent, manger, and soul administrator, Executives director, Principle Investigator PI for FERDT, FERDAT, Fakhraie Bahman Organization, with full rights to take legal action and make final settlement on its accounts.
* Dr. Bahman Fakhraie will act as agent, manger, and soul administrator for Fakhraie Bahman Organization, with full right to take legal action on its accounts.
* A general electronic bookkeeping is followed, on periodical bases, as grants require it. Current standard is quarterly, post full disbursements of grant funding. (FRR is required quarterly)
* Details, information, managerial technologies are legally protected, and no such disclosure will be made of the Organization, administrator, assignees, or representatives of FERDT, FERDAT, or Fakhraie Bahman Organization. All copyright materials are marketable, or further developed by Mr. Fakhraie. All existing rights are reserved.
* All informal and formal requests can be made by mail, or email.

* There is forfeiture of all advances, in cases of mortality, health, or attempt at disbarment, legal action, harm, against soul administrator, Fakhraie Bahman Organization, with full right to take legal action on its accounts.
* Any and all such request will have to forward additional funding for legal representations and undue imposition of costs or harm, please read the legal note and sec. G. It will always apply.
* **Alteration to submit to requirements and regulations will fully transfer all risks, fees, and penalties, and legal responsibilities, to the source of the requirements.**

* The reviewers will have access to a portion of the progress.
* Full transparencies are practiced for authorized agents only, mostly through the internet, after proper signatures are obtained.
* All future development, educational development will have to be contracted, or signed out temporarily.
* All copyrights and proprietary rights are reserved, granted, accorded to Dr. Bahman Fakhraie, and his living trust.
* The final product in initial agreement will be shared with the grantors, after all the financial settlements have been met.

* PI, FERDAT; Dr. Fakhraie, Bahman
Dr. Bahman Fakhraie, PhD, UOU, UT, USA
* Position Classification, Executive level I,
* Salary exceeds Fed Cap for private contracts.
* NSF ID, 000586796, Organizational ID Code, P 269878425
* Private for Profit Business:
* DUNS Number D&B # ███89684,
* CAGE/NCAGE: ███S3, Congressional Districts, UT_001
* Sec G of private contract always applies
* For other institutional DUNS, etc. Please, check cover letter, or contact PI. Thanks
* PI Information(dated)

Grantors information goes here

Section 4, Legal Clause, All Right Reserved & Other Legal Notes
All rights will revert to author after 3 years, if limited rights are contracted.
E. Speech, Lecture Tours, and Research Papers & Follow Up:
All publications or Grant-products are commonly published on completion, USA Library of Congress, Trade Journals, Or Author's websites, etc.
Contact for individualized contracts
F. for Time Donation, And Charity Events: 1. 5% of Net contracts paid
2. Time by Appointments Only.3. Write Donations Checks to Fakhraie Trust fund / Bahman Fakhraie
G. Any and All Risks or Harms invokes these legal clauses. The Following legal clauses are base minimum, and will not limit all rights and legal protection, with uniform constitutional rights they are implied and accorded. Contractually all rights are implied and accorded. Please, read all the legal notes before using any materials.
G1. In Cases of misuse of these materials, plagiarisms, use to put at risk of harm, or to harm, (Legal, or Etc.) the firm, CEO, FERDT,

president: Mr. Bahman Fakhraie, His family members, it will result in $60 million legal action per each event in United States Courts/ any Jurisdictions determined or set by author or a trust set up by Mr. Bahman Fakhraie. Also, use of creative ideas and intellectual properties for any commerce direct or indirect or by proxy, expansions of outlines, attempts to defraud, over charge, harm with illegal acts, misuse of private and confidential materials, information, and properties, letters, checks, e-mails, faxes, any and all communications, and any electronically stored materials and pictures, or false claims against the above entities will be an agreement with and subject to all elements of this contract, and as a user license, without nullifying further and ensuing legal actions or collections, or as directed by the same or authorities. All other contracts are one time contracts and s. t. this contract. Please, notify firm, Mr. Bahman Fakhraie, of all infractions, transactions, or transfer to individuals or to private or government entities in accord with current laws.

G2a. All Costs due to inquiries, related or required business licenses, legal paper works, and hours, all costs related to corrective or reparation publicity, bonding and insurances costs, stock related activities, all costs related to any harm (financial & legal) will be charges to the sources and inquirers, and none to the firm or Bahman Fakhraie, his family, or any and all of their related assets. All attempts to harm, engagements, cashing checks or accepting cash from the same entities is an agreement with all elements of this contract.

G2b. All copy rights associated with written works and further development of all works remains with the author Bahman Fakhraie or his trusts. All development rights are reserved by the same entities.

 G3. Any and all unsolicited commercial phone calls, to Mr. Bahman Fakhraie and, his family will result to a minimum of $ 500 per call charged to the source, due and payable in the same month the call is made, all other elements of this contract also apply, all contracts remain subservient to this contract, all costs of collection and legal actions also are added to the bill payable by source and not Bahman Fakhraie and these entities.

G4.All Consultation and coaching involves risks due to market, business, political, etc. costs and damages related to all such risk are paid by individuals or business entities involved and none is implied or taken by Bahman Fakhraie, his family, or any related assets, under any and all conditions. These entities' court fees are $10,000; et minimum.

 G5.Uncomfortable conditions, misapplications of any law to harm, harassment, odd-hour calls, etc will be due cause for termination of services without refund, and are actionable per this contract.

§§§§

All Right Reserved & Other Legal Notes
All publications or Grant-products are published on completion, USA Library of Congress, Trade Journals, Or Author's websites, etc. All rights revert to author after 3 years, if limited rights are contracted.

E. Speech, Lecture Tours, Research Papers & Follow Up:
1. Per Each Event-$ 350,000 +.
2.   with Foreign Travel-$ 3,000,000
3. Book/Cash Advance floor Min; Net Royalties ~40%
4.   Scripts: /Starting from.....................$ 300,000
5.   Movie Script (120+Pages).................$ 3,000,000
6.   Or Cash Advance Plus 2% of All Gross paid annually
7.   Consultations Fee/Initial charge non-refunded $ 10,000, & ...10 %
8.   Business limited ventures, partnerships, etc. .... ≥ 30%
9.   Salary executive level I

F. for Time Donation, And Charity Events: 1. 5% of Net contracts paid 2. Time by Appointments Only.3. Write Donations Checks to Fakhraie Trust fund / Bahman Fakhraie. G. Any and All Risks or Harms invokes these legal clauses. The Following legal clauses are base minimum, and will not limit all rights and legal protection, with uniform constitutional rights they are implied and accorded. Contractually all rights are implied and accorded. Please, read all the legal notes before using any materials. G1. In Cases of misuse of these materials, plagiarisms, use to put at risk of harm, or to harm, (Legal, or Etc.) the firm, CEO: Mr. Bahman Fakhraie, His family members, will result in $60 million legal action per each event in United States Courts/Jurisdictions determine by author or a trust set up by Mr. Bahman Fakhraie. Also, use of creative ideas intellectual properties, expansions of outlines, attempts to defraud, over charge, harm with illegal acts, misuse of private and confidential materials and properties, letters, e-mails, faxes, any and all communications, and any electronically stored materials and pictures, or false claims against the above entities will be an agreement with all elements of this contract, and as a use license, without nullifying ensuing legal actions or collections, or as directed by the same or authorities. Please, notify firm, Mr. Bahman Fakhraie, of all infractions, transactions, or transfer to individuals or to private or government entities. G2a. All Costs due to inquiries, related or required business licenses, legal paper works and hours, all costs related to corrective or reparation publicity, bonding and insurances costs, stock related activities, all costs related to any harm (financial & legal) will be charges to the sources and inquirers, and none to the firm or Bahman Fakhraie, his family, or any and all of their related assets. All attempts to harm, engagements, cashing checks or accepting cash from the same entities is an agreement with all elements of this contract. G2b. All copy rights associated with written works and further development of all works remains with the author Bahman Fakhraie or his trusts. Any and all development rights are reserved by the same entities. G3. Any and all unsolicited commercial phone calls to Mr. Bahman Fakhraie, his family will result to a minimum of $ 500 per call charged to the source, due and payable in the same month the call is made, all other elements of this contract also apply, all contracts remain subservient to this contract, all costs of collection and legal actions also are added to the bill payable by source and not Bahman Fakhraie and these entities. G4.All Consultation and coaching involves risks due to market, business, political, etc. costs and damages related to all such risk are paid by individuals or business entities involved and none is implied or taken by Bahman Fakhraie, his family, or any related assets, under any and all conditions. These entities' court fees are $10,000; et minimum. G5.Uncomfortable conditions, misapplications of any law to harm, harassment, odd-hour calls, etc. will be due cause for termination of services without refund, and are actionable per this contract.

§§§§

Letter of Agreement with FRDET and Fakhraie Bahman

I/We (Ms., Mrs., Mr.):   ....

Business Name: ....

Social Security numbers: ..../......./............ ..../......./................

Tax ID Numbers: ....

Driver License Numbers:

....

Passport (Birth Certificate) Numbers:

....

Have read all the information supplied by Dr. Bahman Fakhraie, and /or FRDET (Trust, LLP). Agree to supply all correct financial and related information and select the following services by marking (X), or Letter a b c, Or Number 1 2 3, or write and specify. Agree to the power

of attorney required to confirm or acquire financial and related
information. Plus a check for $3000, for preliminary registrations,
legal and financial inquiries, and licensing fees.
   1: ☐   2: ☐   3: ☐   4: ☐   5: ☐   6:
☐   7: ☐   8: ☐
Specify which kinds of the following accounts you will require:
Individual/Gov. ☐   Joints ☐   Joint with
survival rights ☐   Trusts   ☐
        Details: ....
Business types:
    Ltd Partnership (LLP, LLC) ☐   Soul Owner ☐
    Inc/Gov ☐   Firm ☐
        Details: ....

I/We agree to pay the fees and costs as they accrue, after 30 days a
charge of %10 per annum is added to past due amounts until they are
settled or paid in full. That at least 75% charges and expenses are paid
no later than six month after initial reports or outlines are examined.
That all travel lodgings and phone charges, legal research, and research
hours, I/We request are fully paid. I/We have read and understand the
nature of business, business cycles, market price fluctuation risks,
currency fluctuation risks, social and political risks, natural and
climatic risks, and all other risks herewith not itemized; therefore, I/we
accept all financial losses, responsibilities, all punitive or compensatory
damages that occur for all activities that are undertaken based on or
claimed related to the report or reports, advise and etc., generated by
Money Wise Firm, Bahman Fakhraie or all the entities named, agents,
and associates. I/We understand in cases of any false, misleading, or
withholding information the Money Wise Firm and all entities named
above will not be held accountable, responsible, financially or
otherwise, and will be refunded for any damages fully. I/We release all
others associated with Money Wise Firm, and Bahman Fakhraie from
financial losses, responsibilities, all compensatory and punitive charges
concerning activities I/We undertake.
Sign (full name, titles, and address)
 Business (Name, Address; Agents' title):

Section 5, Organization Chart

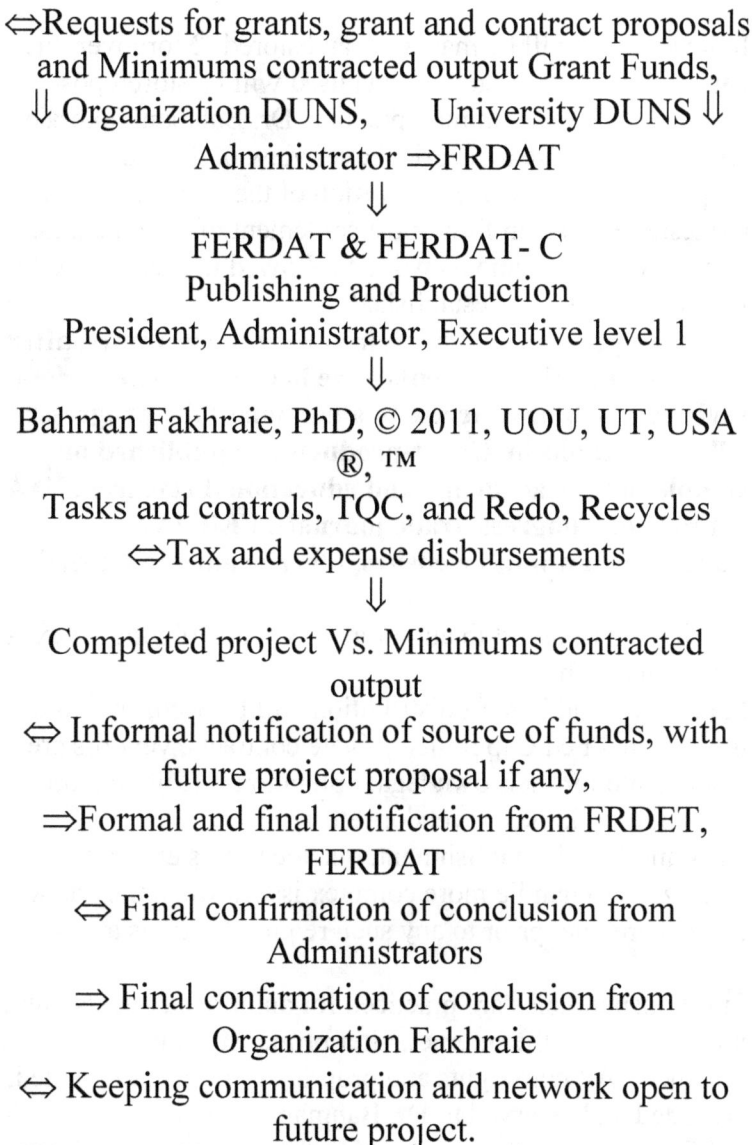

⇔Requests for grants, grant and contract proposals and Minimums contracted output Grant Funds, ⇓ Organization DUNS, University DUNS ⇓ Administrator ⇒FRDAT
⇓
FERDAT & FERDAT- C
Publishing and Production
President, Administrator, Executive level 1
⇓
Bahman Fakhraie, PhD, © 2011, UOU, UT, USA
®, ™
Tasks and controls, TQC, and Redo, Recycles
⇔Tax and expense disbursements
⇓
Completed project Vs. Minimums contracted output
⇔ Informal notification of source of funds, with future project proposal if any,
⇒Formal and final notification from FRDET, FERDAT
⇔ Final confirmation of conclusion from Administrators
⇒ Final confirmation of conclusion from Organization Fakhraie
⇔ Keeping communication and network open to future project.

Bahman Fakhraie, PhD, © ® ™ 2010-, UOU, UT, USA

Section 6, DATA MANAGEMENT:

Relevant data will be managed, and stored. Moreover, the data and the progression of data used will be stored post completion. As a common practice, Dr. Bahman Fakhraie will, also preserve a historical over view of data for legal purposes, for future use. A portion of that will be shared if requested in written form, post settlement of all expenses. The ethical requirements will be followed in accordance with the applied law established.

The final product will be shared with the grantors, after all the financial settlements have been met. The reviewers will have access to a portion of the progress. All publications or Grant-products are published on completion for academic and educational research, USA Library of Congress, trade journals, FERDAT Publishing, author's websites, other academic journals, etc.

All rights will revert to author after 3 years, if contract exits for limited rights.

Labor laws and benefit distribution will be according to the institutional Fed Cap limits, private contracts; vendors are responsible for their own legal obligations. (Publication Ink, FedExx, kinkcos, vendors etc.)

CPA and Legal establishment, outside to this entity will be contracted to handle more complex issues when they arise, all cost are due, prior to any such required actions are requested.

Full transparencies are practiced for authorized agents only, mostly through the internet, after proper signatures are obtained. All copyrights and proprietary rights are granted, accorded and reserved to Dr. Bahman Fakhraie,

All future development, educational development will have to be contracted, or signed out temporarily, they will revert to Dr. Bahman Fakhraie, in case of all legal issues.

These and other clauses will be amended and upgraded as required over time, they all apply.

Absolut or required non-disclosures by the firm or Dr. Fakhraie will require pre-paid plan with minimum flooring pay. Flex-legalism targeted at the firm, or Dr. B. Fakhraie will be due cause for nonpaid termination by the same.

G. Any and All Risks or Harms invokes these legal clauses. The Following legal clauses are base minimum, and will not limit all rights and legal protection, with uniform constitutional rights they are implied and accorded. Contractually all rights are implied and accorded. Please, read all the legal notes before using any materials.

Sec 7, Check CURRICULUM VITA

SELECTED BIBLIOGRAPHY

Bailey, Martin J., *National Income and the Price Level: A Study in Macroeconomic Theory,* New York, McGraw-Hill, 1971.

Baumol, William J., *Economic Theory and Operational Analysis,* 4th ed., New Jersey: Prentice-Hall, 1977.

Blaug, Mark, *Economic Theory in Retrospect,* 3rd Ed., London: The Cambridge University Press, 1978.

Böhn_Bawerk, Eugen Von., *Capital and Interest: Positive Theory of Capital,* vol. II, Trans. G. D. Hunt & H. F. Sennholz, Chicago: Libertarian Press, 1959.

Boulding, Kenneth E., *Economics As A science*, New York: McGraw-Hill Book Company, 1970.

Byers, Lloyd L., *Concept of Strategic Management: Planning and Implementation*, New York: Harper & Row Publishers.

Caves, Richard E. and Jones, R. W., *World Trade and Payments: An Introduction*, 2nd ed., Boston, Little, Brown and Company, 1977.

Chiang, Alpha C., *Fundamental Methods of Mathematical Economics*, 2nd ed., New York: McGraw-Hill Book Company, 1974.

Coombs, Philip H., *The World Educational Crisis: A System Analysis*, London: Oxford University Press, 1968.

Domar, Evsey D., *Essays in the Theory of Economic Growth*, New York: Oxford University Press, 1957, 154-167, 168, 181.

Druker, Peter F., *Innovation and Entrepreneurship*, New York: Harper & Row Publishers, 1985.

Eisner, R. "Depreciation Allowances, Replacement Requirement, and Growth," *The American Economic Review*, XLII, December 1952.

Dr. Bahman Fakhraie, *TECONOMIC OF VERBALISM*, Utah, FERDAT publishing 2012, and
Paperback link is at, https://www.createspace.com/4121720
The EBook Link is at,
http://www.amazon.com/dp/B00B1LO7UQ
Books web page at, http://bahfecon.wix.com/bahfecon#

—. The Demand and Supply Sides of Appropriate Technological Advancement, (Research paper at University of Utah Economics Dept. 2003)

—. *Demand and supply Sides of Technological Injections*, Utah, FERDAT Publishing, 2004, And at,
http://www.amazon.com/dp/098529583X/ref=rdr_ext_tmb
#reader_098529583X

—. *Technological injection, dynamic new capital measurements, and Production Theory in Economics*, (Michigan: ProQuest LLC, 2010) and,
https://order.proquest.com/OA_HTML/pqdtibeCCtpItmDsp Rte.jsp
Books web page at, http://bahfecon.wix.com/bahfecon#

—. *Teconomics of Verbalism*, Utah FERDAT Publishing 2012, and at, https://www.createspace.com/4121720, Books web page at, http://bahfecon.wix.com/bahfecon#

—. *Analytical Remedies for The Millennial Cascading Economic Declines*, Utah FERDAT Publishing 2012, and at, https://www.createspace.com/4187823, Books web page at, http://bahfecon.wix.com/bahfecon

—. "Economic Theories and Practices in Technological Changes, capital measures, and Production." (Research paper at University of Utah Economics Dept. 1988)

—. "Hallowing headless nations: the need to invest on public education under the 1980s international economic conditions," (Research paper at University of Utah Economics Dept. 1988)

—. "Transfer of Technologies and Socioeconomic Theories of Dualism," (Research paper at University of Utah Economics Dept. 1983)

Friedman, Milton and Schwartz, Anna Jacobson, *The Great Contraction 1929-1933*, vol.2, 2nd ed., New Jersey: Princeton University Press, 1973

Gander, James Patrick., *Technological Change and Raw Materials*, Salt Lake City: Bureau of Economic and Business Research, University of Utah, 1977.

Gardner, Ackley., *Macroeconomics: Theory and Policy*, New York: Macmillan Publishing Co., 1978.

Girton, Lance and Roper Don, "Theory and Implication of Currency Substitution," *Journal of Money, Credit, and Banking*, 13, no. 1 (February 1981): 12-30.

Hayek, Friedrich August Von. , "Kapitalaufzehrung." Weltwirtschaftliches Archive 36, 1932, II, 86-108.

Heilbroner, Robert L., *The Worldly Philosophers*, New York: Time Inc., Special Ed., 1962.

Hicks, Sir John, *The Crisis in Keynesian Economics*, New York: Basic Books, Inc. 1974.

Hirshleifer, Jack, *Investment, Interest, and Capital*, New Jersey: Prentice-Hall, 1970.

—. *Time, Uncertainty, and Information*, New York: Basil Blackwell, 1989.

—. *Price Theory and Applications*, New Jersey: Prentice-Hall, 1976.

Henderson, J. M. and Quandt, R. E., *Microeconomic Theory; A Mathematical Approach*, 2nd ed., New York: McGraw-Hill, 1972. 191-199, 280

Hunt, E. K. and Howard J. Sherman, *Economics: An Introduction to Traditional and Radical Views*, 2nd Ed., San Francisco: Harper and Row Publishers, 1975.

Intriligator, Michael D., *Econometric Models, Techniques, and Applications*, New Jersey: Prentice-Hall, 1978.

Johnson, J., *Econometric Methods*. New York: McGraw-Hill, 1972.

Kant, Immanuel, *Critique of Judgment*, Trans. J. H. Bernard, New York: Hafner, 1951.

Kennedy Charles and Thirlwall, A.P., "Surveys in Applied Economics: Technological Progress," *The Economic Journal*, March 1972: 12.

Keynes, John Maynard, *Essays in Biography*, London: The Cambridge University Printing House for Royal Economic Society, 1972.

—. *The General Theory of Employment, Interest and Money,* 1st ed. 1936, London: The Cambridge University Printing House for Royal Economic Society, Reprint 1973.

Kindleberger, Charles P., *The World in Depression 1929-1939*, Los Angeles: University of California Press, 1973.

Kirzner, I.M., *Discovery and the Capitalist Process*, Chicago: University of Chicago Press, 1985.

Klein, P. A. and Moore, G. H., *Monitoring Growth Cycles in Market-Oriented Countries*, (Mass.: published for N. B. E. R., by Ballinger publishing Co., 1985).

Knight, Farnk H., *Risk, Uncertainty and Profit*, Chicago: University of Chicago Press, 1985.

Landes, David S., *The wealth and poverty of nations: Why Some are So Rich and Some are So Poor*, (New York: W.W. Norton & Company, 1998).

Lindert, Peter H. and Kindleberger Charles P., *International Economics,* 7th Ed., Illinois: Richard Irwin, 1982.

Mansfield, Edwin, *Technological Change*, New York: W. W. Norton & Co., 1971.

Mark, Blaug, *Economic Theory in Retrospect*, 3rd Ed. London: The Cambridge University Press, 1978.

Marx, K. (In a letter he wrote to Engel, dated 20 August, 1862, London), (handout by Professor Randa, 2004)

McKinnon, Ronald I., *Money and Capital in Economic Development*, Washington D. C.: The Brookings Institution, 1973.

Mensch, Gerhard, Das *technilogische Patt.*, Frankfurt: Umschau Verlag, 1975.

—, *Stalemate in Technology: innovation overcome the Depression*, (Massachusetts: Ballinger Publishing Company, 1979).

Miles, T.R., "Gestalt Theory," in The Encyclopedia of Philosophy, New York: Macmillan Publishing Co., vols. 3 and 4, 1967.

Mishan, Edward Joshua, *Cost-Benefit Analysis,* 4th Ed. London: Unwin Hyman, 1988.

Moran, Michael. "New England Transcendentalism," in The Encyclopedia of Philosophy, (New York: Macmillan Publishing Co., Vols. 3 and 4, 1967).

Mundell, Robert A. , "Growth, Stability, and Inflationary Finance," Journal *of Political Economy*, 73, 1963.

Pindyck, R.S., and Rubinfeld, D.L., *Econometric Models and Econometric*, 2nd ed., New York: McGraw-Hills, 1981.

Ott, J. Steven, *The Organizational Culture Perspectives*, (Pacific Grove, California: Brooks/Cole Publishing Company, 1989).

Quirk, James and Saposnik, Rubin, *Introduction to General Equilibrium Theory and Welfare Economics*, New York: McGraw-Hill, 1968.

Rima, Ingrid, *Development of Economic Analysis*, 7th ed., New York: Routledge, 2009.

Rodinson, M., *Islam and Capitalism*, Trans. B. Pearce, Austin: University of Texas, 1981.

Rosenberg, Nathan, Technology *and American Economic Growth*, New York: M. E. Sharp, 1972.

Ruttan, V. W. , "Usher and Schumpeter on Invention, and Technological Change," Quarterly *Journal of Economics*, 1960, 602

Ruttan, Vernon W. , "Usher and Schumpeter on Invention, Innovation, and Technological change," Quarterly *Journal of Economics*, 1960, 73

Samuelson, Anthony Paul. Economics, 11th Ed., New York: McGraw-Hill, 1980.

Savich, R.S., and Thomson, L. A., *Resource Allocation within the Product Life Cycles*, Business Topic, MSU: MSU, fall 1978.

Schmooker, Jacob, *Invention and Economic Growth*, Massachusetts: Harvard University Press, 1966.

Schultz, Theodore W., *Investing in People: The Economics of Population Quality*, Berkley: University of California Press, 1982.

Schumpeter, Joseph A., *Business Cycles*, vol. 1. New York: McGraw Hill Book Company, 1939.

—. *Business Cycles: A Theoretical, Historical, and Statistical Analysis of the Capitalists Process,* vol. I, New York, McGraw Hill Book Company, 1938.

—. *The Theory of Economic Development: An Inquiry into Profits, Capital, Credit, Interest, and Business Cycle*, Trans., Redvers Opie, London: Oxford University Press, Reprint 1980.

Smith, Adam., *An Inquiry into the Nature and Cause of Wealth of Nations*, Edited by E. Cannon, Chicago: University of Chicago Press, 1976.

—. *An Inquiry into the Nature and Causes of Wealth of Nations*, vol. 2, 2nd ed., Oxford: the Clarendon Press, 1988.

—. *The Theory of Moral Sentiments*, London, 1st ed., 1757.

Smith, E.J. Chambers, R.H. Scott and R.S., *National Income Analysis and Forecasting*, Glenview: Scott, Foresman and Company, 1975.

Spiegle, Henry William., *The Growth of Economic thought*, North Carolina: Duke University Press, 1983.

Stigler, George J., *The Theory of Price*, 3rd ed., New York: The Macmillan Company, 1966.

Takaki, Ronald., *A Different Mirror: A History of Multicultural America*, London: Little Brown and Company, 1993.

Taylor, John R., *An Introduction to Error Analysis*, Mill Valley: University Science Books, 1982.

Thirtle, Colin G. and Ruttan V. W., *The Role of Demand and Supply in the Generation of Diffusion of Technological Change*, Switzerland, 1987.

U. S. President, *Economic Report of the President*, (Washington, D.C.: Government printing office, 1990)

Usher, Abbot Payson, *A History of Mechanical Inventions*, London: Oxford University Press, 1954.

Usher, Abbot Payson, *A History of Mechanical Inventions*, Revised ed., London: Oxford University Press, 1954.

Viner, Jacob, *Studies in the Theories of International Trade*, New York: Harper and Brothers publishers, 1937.

Wainwright, A. C. Chiang and K., *Fundamental Methods of Mathematical Economics*, 4th ed., Boston: McGraw-Hill Irwin, 2005.

Webster, Merriam, *Merriam-Webster's Collegiate Dictionary*, 9th ed., Springfield: M.W. Inc., 1985.

Young, Hugh D., *Statistical Treatment of Experimental Data*, New York: McGraw-Hill, 1962.

CURRICULUM VITA

Bahman Fakhraie, PhD, UOU, UT, USA
University Of Utah, Economic Dept., U. O. U.;   Salt Lake
City, Utah USA, 84112
Permanent Address: c/o 1120 Canyon Rd No. 29; Ogden
Utah 84404
E-Mail: bf9@utah.edu , bahf.econ@gmail.com,
dr.bahf.econ@gmail.com
Dr. Bahman Fakhraie's Books webpage,
http://bahfecon.wix.com/bahfecon
Honors and Awards: **Omicron Delta Epsilon Honor
Society**
Utah State U. USA, **Certificates Keys to Agricultural
Development at the Local Level**
Student Leadership Positions: **President of International
Student Association** [ASUSU]
Bachelor of Science: **Utah State University**
Master of Science: **Utah State University**
PhD, University of Utah   **University of Utah**
PhD, Economics (international economics), University of
Utah
**Certificate of Completion, PhD in Economics University
of Utah, 2010**
**Certificate of Completion, PhD in Economics University
of Chicago, 2011**
<u>Mission Statement:</u>
To formulate optimum functionality in management teams,
in order to make profitable and beneficial contributions,
and create opportunity for merited advancements.
Utilizing rare combination of theory and empiricism,
cultivated in multi-cultural inclusive settings in academic
and private enterprise, I enhance the productivity and
completion of project management. Experiences, among
fields of International Trade, Managerial, and Production
theories in Economics, Finance, and Business, which

modern business and academic institutions find very expensive to employ and too costly not to employ. The extensive background helps advance research in my field of study.

Goals:

I have initiated, passed, and funded many constructive projects --goals by committees--, where it has been value enhancing and mutually beneficial individually, and by team assists. Private Senior Economist: Research Positions with Contracts, Teaching, Research, Books, with contract.

Short Term Private Contracts are for economic and financial educational consultations.

Academic Preparatory Continuum:

PhD, University of Utah, Dissertation:

TECHNOLOGICAL INJECTION, DYNAMIC NEW CAPITAL MEASUREMENTS AND PRODUCTION THEORY IN ECONOMICS

Thesis statement:

The dynamic influences of technology and elemental factors of production --defined and measured in this dissertation-- are greater than commonly have been calculated or expected. The impacts of different measurements of capital stocks (traditional and new adjusted capital) on embodied and disembodied technological variables, on productivity, and economic growth of national social products are tested. The econometric effects of two new capital stock measures introduced in the writing of Friedrich August Von Hayek, John Maynard Keynes, and further developed by Evsey D. Domar, and ignored by most modern economists are examined. Therefore, we focus on the demand and supply side of technological embodied in capital, in human skill and produced goods and services, and the economy. Dr. Bahman Fakhraie's Book-link is at, https://order.proquest.com/OA_HTML/pqdtibeCCtpItmDspRte.jsp

Post-Doctoral Research and Goals:
1. Updating the econometrics of the dissertation to most recent available data and dates, (per available grants)
2. Include, countries, with acceptable data in the study, set up formulaic development in excel etc. for use. (per grants)
3. Focus on the theoretical advancement in production processes for systemic innovational additive methodologies, in Movie production, Agriculture, and other creative industries, using existing or newly hired faculties, in anchored and linked institutions. This is a great serious work in shadows of Adam Smith, Schumpeter, Hayak, and Keynes.

    Teconomics: Scientific Synthesis of Economics and Technology in Teconomics

4. Stream line methodologies, and project for PhD Students, master students, and introductory fields' level, an extension of the dissertation and current copyrighted writings.
5. A multimedia production of Dr. Fakhraie's recent research,
NEW DYNAMIC ECONOMIC MODELS TO STUDY TECHNOLOGY INJECTIONS & DYNAMIC CAPITAL FORMATION, IMPACTS ON INTERNATIONAL TRADE, EXCHANGE RATE, AND GLOBAL ECONOMIC OPTIMASATION

    Presentation by Dr. Bahman Fakhraie

6. Of course, this will be under Sec G of private contract and without infringing on rights of marketing, publishing, and distribution of the same. Certain contracts will be more restrictive, including specific names.
7. Finish the books series in post dissertation Teconomic studies, Micro, Macro, Teconometrics,

Teconomic Analysis, and Political Teconomics. (in progress)

Books, published: DR. Bahman Fakhraie's Books web page at, http://bahfecon.wix.com/bahfecon

Dr. Bahman Fakhraie, *TECONOMIC OF VERBALISM,* Utah, FERDAT publishing 2012, a
Paperback link is at, https://www.createspace.com/4121720
The EBook Link is at,
http://www.amazon.com/dp/B00B1LO7UQ
Books web page at, http://bahfecon.wix.com/bahfecon

- *Demand and supply Sides of Technological Injections*, Utah, FERDAT Publishing, 2004, And at,
  http://www.amazon.com/dp/098529583X/ref=rdr_ext_tmb reader_098529583X

- *Teconomics: the microeconomic analysis*, Utah, FERDAT publishing 2012, and,
  https://www.createspace.com/4196760, Books web page at,
  http://bahfecon.wix.com/bahfecon

- *Technological injection, dynamic new capital measurements, and Production Theory in Economics*, (Michigan: ProQuest LLC, 2010) and,
  https://order.proquest.com/OA_HTML/pqdtibeCCtpItmDspRte.jsp
  Books web page at,
  http://bahfecon.wix.com/bahfecon

- *Teconomics of Verbalism*, Utah FERDAT Publishing 2012, and at,
  https://www.createspace.com/4121720,

Books web page at,
http://bahfecon.wix.com/bahfecon

- *Analytical Remedies for The Millennial Cascading Economic Declines,* Utah FERDAT Publishing 2012, and at,
  https://www.createspace.com/4187823,

- "The Demand and Supply Sides of Appropriate Technological Advancement." (Research paper at University of Utah Economics Dept. 2003),

- "Economic Theories and Practices in Technological Changes, capital measures, and Production." (Research paper at University of Utah Economics Dept. 1988).
- Fakhraie, Bahman, "The Demand and Supply Sides of Appropriate Technological Advancement." (Research paper at University of Utah Economics Dept. 2003).
- "Hallowing headless nations? The need to invest on public education under the 1980s international economic conditions," (Research paper at University of Utah Economics Dept. 1988).
- "Transfer of Technologies and Socioeconomic Theories of Dualism," (Research paper at University of Utah Economics Dept. 1983).

Production theory in Agriculture
Agro-production, Sheep production in Iran (an onsite research project 1975)
Monetary Macroeconomic Specialization (Milton Freedman,)
International Monetary macroeconomic Specializations (Robert Mundell)
Fiscal Analysis of oscillatory modifications
Applied Agricultural production in developing economies

Inappropriate Technology Transfers by Corporations and dualistic induced instabilities in a pre-democratic economies.

Certificate in Rural Development from Utah State University

Research Skills:

Statistical packages, data selection, analysis, and formulations.

Computer Languages, Spread Sheet Data Analysis, Writings

Multimedia & Creative Skills: Film and video production, CD: audio and video works

Languages: Fluent in English and Farsi, I can read and write some French, and Arabic.
                    Fluent in Dezfili Dari dialectic [One of the earliest spoken languages]

Teaching Experiences:

University of Utah: macroeconomic and microeconomic: introductory course, and related mathematics.

Utah State University: Economic Department, Persian language, and cultural studies to faculty and students, and teaching assistance for Agricultural economics.

Volunteer Helping of other students

Student leadership positions AUSU, as an undergraduate

PROFESSIONAL EXPERIENCE:

Founder and Management of Money Wise Firm (A Private Financial Education Foundation for Personal Financial studies), 1970 to date

Volunteer work: President; V.P., and Treasurer of Cherrywood Association Inc., multimillion dollars project, (Different 3 year cycles 1980 to 1999, consultation to date)

Real Estate Investor and Renovations, General Manger, to date

Agribusiness management, owner manager

Auto Agency sale manager/sold

Auto shop management/sold

Numerous Volunteer Projects: planning, budgeting, contracting, and finishing.

Beside educational publications, and digital multi-media CDS, DVD formats, copyrighted at USA the library of congress.

Other Experience:

Budget Analysis, Budget Setting, Budget Forecasting, Asset Allocation Studies, Portfolio Analysis Studies, Saving (Goal Setting) Plan, Tax Management Studies, Risk/Reward Management and studies, Economic Condition risk Analysis, Managerial Goal Setting and Project Production Process and Enactments.

Economic Topics of Research and Lectures:

Non-Marxist Revolutions in Middle East (Iran)

Economic and Indices (Measurement Issues)

Economic Theories of Technological Changes and Capital Measures

Technological Change, Growth Rate, and Capital Formations with US Data

Education and Taxation

Hollowing Headless Nations! Education Crisis

Technological Parameters Statistical Measurements

New Econometric Measurements of Capital in Production Theory

Reading (Economic, Science, Mathematics, Econometrics, Statistics, Mystery, Ancients), Films (art, industry), Jazz, Foreign Eclectics music, Fly-fishing

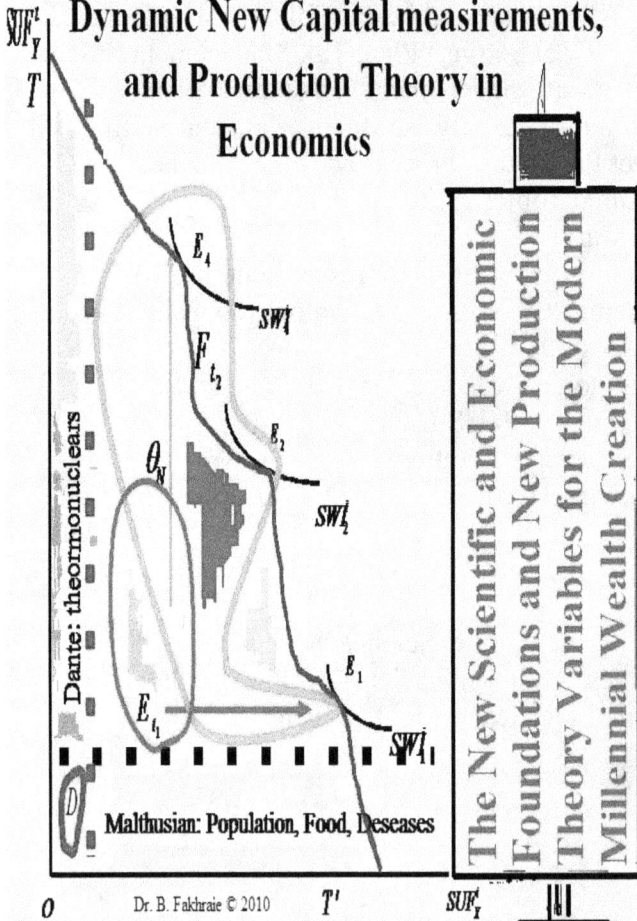

Techonological Injection, Dynamic New Capital measirements, and Production Theory in Economics

The New Scientific and Economic Foundations and New Production Theory Variables for the Modern Millennial Wealth Creation

DEMAND & SUPPLY SIDES OF TECHNOLOGICAL INJECTIONS

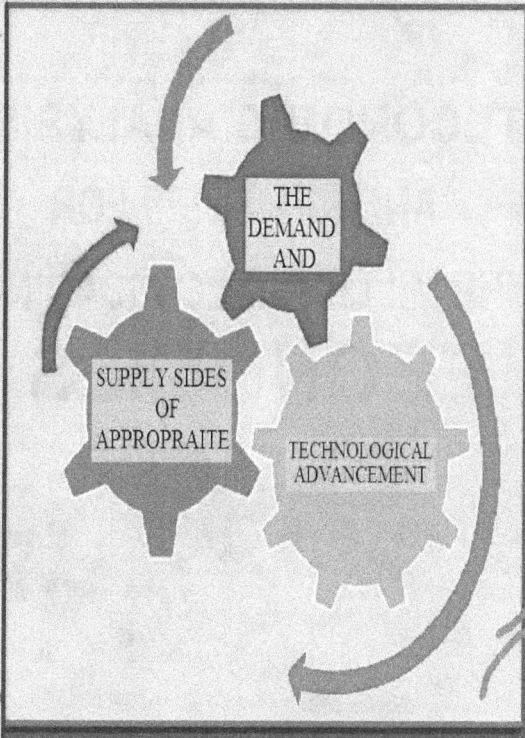

Supply

Demand

Technological
injections

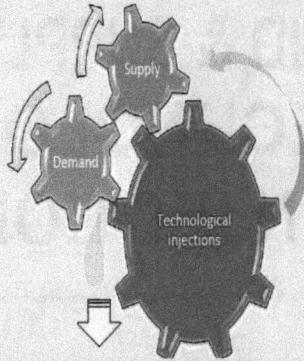

TECONOMIC ANALYSIS
AND REMEDIES FOR
THE MILLENNIALCASCADING
ECONOMIES

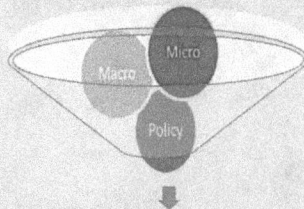

Micro

Macro

Policy

Solutions, beside repeating past mistakes.

Striped Subset: Trade surplus sustainable optimization possible

Environmental Factors:
1-Lethaly toxic
2-Non toxic

Doted Subset:
Profit maximization
porbabale

TECONOMICS

Economics:
1-Relative competetion
2-Non competetive

Political Conditions:
1-Relative democracy
2-Non-demcratic

SCIENTIFIC SYNTHESIS OF MICROECONOMICS
AND TECHNOLOGICAL INJECTIONS
IN ECONOMICS

Global Trade:
1-Free trade
2-Limted trade

Dr. Bahman Fakhraie © 2010

THE MILLENNIAL POLITICAL ECONOMIC PARADIGM

Dr. Bahman Fakhraie's books,
http://bahfecon.wix.com/bahfecon

Dr. Bahman Fakhraie, PhD in Economics, University of Utah, and his dissertation added to the influences of the Unorthodox Holistic Economic doctrine and complemented the modern orthodox economic theories, in the millennial age of technological paradigm shifts. He applies analytical skills with gestalt study of history, mathematics, and econometrics to economic analysis, with scientific background. He is a Published Economist, Author, Researcher, Investor, and Private Contractor. His skills are in international trade and finance, economic production (theory and application), growth and development theory, econometrics, agriculture economics, and agronomy. These are greatly valued skills combinations to employ.

Books' webpage, http://bahfecon.wix.com/bahfecon

Library of Congress copyright 1851630
ISBN-10: 0985295856, ISBN-13: 978-0-9852958-5-1

$28.95

ISBN 978-0-9852958-5-1

52895>

9 780985 295851